S0-DUW-105

HISTORY OF BROADCASTING: RADIO TO TELEVISION

HISTORY OF BROADCASTING: Radio to Television

A History of Wireless Telegraphy

J. J. FAHIE

ARNO PRESS and THE NEW YORK TIMES

New York • 1971

Reprint Edition 1971 by Arno Press Inc.

LC# 77-161144
ISBN 0-405-03565-9

HISTORY OF BROADCASTING: RADIO TO TELEVISION
ISBN for complete set: 0-405-03555-1
See last pages of this volume for titles.

Manufactured in the United States of America

A HISTORY

OF

WIRELESS TELEGRAPHY

Printed by

WILLIAM BLACKWOOD & SONS, *Edinburgh, Scotland.*

THE ARCH-BUILDERS OF WIRELESS TELEGRAPHY.

A HISTORY

OF

WIRELESS TELEGRAPHY

INCLUDING SOME BARE-WIRE PROPOSALS
FOR SUBAQUEOUS TELEGRAPHS

BY

J. J. FAHIE

MEMBER OF THE INSTITUTION OF ELECTRICAL ENGINEERS, LONDON, AND OF
THE SOCIÉTÉ INTERNATIONALE DES ÉLECTRICIENS, PARIS;
AUTHOR OF
'A HISTORY OF ELECTRIC TELEGRAPHY TO THE YEAR 1837,' ETC.

WITH FRONTISPIECE AND ILLUSTRATIONS

SECOND EDITION, REVISED

NEW YORK
DODD, MEAD, AND CO.
EDINBURGH AND LONDON
WILLIAM BLACKWOOD AND SONS
MCMI

"Every student of science should be an antiquary in his subject."—CLERK-MAXWELL.

Dedicated

TO

SIR WILLIAM H. PREECE, K.C.B., F.R.S.,

Past President, Institution of Electrical Engineers;
President, Institution of Civil Engineers,
&c., &c.,
The First Constructor of a Practical
Wireless Telegraph,

AS A SLIGHT TOKEN OF ESTEEM AND FRIENDSHIP,

AND IN ACKNOWLEDGMENT OF MANY

KINDNESSES

EXTENDING OVER MANY YEARS.

PREFACE TO SECOND EDITION.

FROM the fact that two impressions of this work have been sold out in fifteen months and that a second edition is now called for, the author is glad to think that he has met a want, and, judging by the press notices, has met it in a satisfactory manner.

While acknowledging with thanks the numerous and, with one exception, altogether favourable reviews of his book, the author begs leave to notice two objections which have been advanced by more than one of his critics.

Firstly, it has been thought that a history of wireless telegraphy now is premature—that the subject is still in a more or less embryonic, or at least infantile, stage, and that the time for writing its history has not yet come. But a beginning has to be made at some time, and as well now as later, and for this reason: While (as stated in the Preface to the first edition) the book is intended to be a popular account of the origin and progress of the subject, the author thought that it would also be useful to students and inventors, as showing them what has been, so far, done or attempted, so that they may not waste their ingenuity on ways and means that have already been exploited.

Secondly, it has been objected that there is much in the book—especially in the First Period—that might be omitted, or still further condensed. But here again the author had in view the requirements of the inventive reader, for whom the crudities and failures of previous experimenters are in their way as instructive as their successes

In this new edition he has made some alterations and additions (chiefly in the pages dealing with the Marconi system), with a view of (1) correcting inaccuracies of expression in some places, and making the meaning more clear in others; (2) bringing out more some points of the theory and practice of Hertzian-wave telegraphy; and (3) bringing up to date the record of Mr Marconi's public demonstrations.

A new and fuller index is appended, in which every subject is noted both under the authors' names and under the subjects themselves. This should make easy the reader's search for any matter that may specially interest him.

In the way of practical applications of wireless telegraphy since the first edition was published in October 1899, Sir William Preece's system has found new employment, as mentioned at p. 160. As regards the Hertzian-wave form, we have many new experimenters in the field, whose "inventions," although generally said to be unlike Marconi's, seem to differ from it chiefly in points of constructive detail; many new demonstrations by Marconi and his imitators of the value of their system or systems, which, within limits, nobody contests; many paragraphs in the newspapers as to what each one is going to do; but so far as actual installations under the rough-and-tumble conditions of everyday working, it must be confessed that progress has been slow—disappointingly so to some people, Sir William Preece, for instance, who is "getting tired of wireless telegraphy," and asks "where is there at present a single circuit worked commercially on a practical system of wireless telegraphy?"

Well, the position is not so bad as Sir William would have us infer. To begin with, there is no longer any question of its value to governments for naval and military purposes, or of its commercial value for outlying islands, lightships, lighthouses, and for shipping generally. To have thus convinced the general public, in the short period of four years, of the soundness of its scientific basis and of its practical utility is no slight achievement, and is all in the way of progress. Then, as a matter of fact, Mr Marconi's system, or some modification of it, *has* been adopted in the navies of all the great Powers, and on some German and Belgian trading vessels. That it has not yet been employed on British vessels of the same kind is not entirely Mr Marconi's fault, but seems more to be due to official obstacles.

Then again, in May of last year, the Marconi apparatus was installed at Borkum, Germany, on a semi-commercial basis ('Electrician,' July 20, p. 488), and about the same time it was introduced into Hawaii as a permanent means of intercommunication between the five islands of the group

('Electrician,' March 2, p. 680). Quite recently a Marconi station has been established at La Panne (Belgium), between Ostend and Dunkirk, and about 61 miles from Dover. The Princess Clementine, one of the Belgian mail packets running between Ostend and Dover, has also been fitted up, and keeps up communication with La Panne in her daily trips across channel. Not only this, but wireless messages have been exchanged between the ship at Dover and the Marconi station at Dovercourt, near Harwich, a distance of over 80 miles of sea and land (London daily papers, November 5-10). Progress, therefore, there has been—slow perhaps, but solid and, all things considered, satisfactory. And now we seem to be on the eve of further extensions, as to which those interested will find some indication in the addresses of the Marconi Company's chairman reported in the 'Electrician,' March 2, August 3, and December 21 of last year.

January 1901.

PREFACE TO FIRST EDITION

EARLY in 1897 there was a great flutter in the dove-cotes of telegraphy, and holders of the many millions of telegraph securities, and those interested in the allied industries, began to be alarmed for the safety of their property. Mysterious paragraphs about the New, Wireless, or Space Telegraphy, as it was variously called, kept appearing in the papers; and the electrical profession itself—certainly some leading members of it—seemed disposed to accept implicitly the new marvels, without the grain of salt usual and proper on such occasions.

In a lecture on Submarine Telegraphy at the Imperial Institute (February 15, 1897), Professor Ayrton said: "I have told you about the past and about the present. What about the future? Well, there is no doubt the day will come, maybe when you and I are forgotten, when copper wires, gutta-percha coverings, and iron sheathings will be relegated to the Museum of Antiquities. Then, when a person wants to telegraph to a friend, he knows not where, he will call in an electro-magnetic voice, which will be heard loud by him who has the electro-magnetic ear, but will be silent to every one else. He will call, 'Where are you?' and the reply will come, 'I am at the bottom of the coal-mine,' or 'Crossing the Andes,' or 'In the middle of the Pacific'; or perhaps no reply will come at all, and he may then conclude the friend is dead."

Soon after, in the course of a debate in the House of Commons (April 2, 1897) on the Telephone monopoly, one of the speakers said: "It would be unwise on the part of the Post Office to enter into any very large undertakings in respect of laying down telephone wires until they had ascertained what was likely to be the result of the Röntgen form of telegraph, which, if successful, would revolutionise our telephonic and telegraphic systems."

When cautious men of science spoke, or should I not say dreamt thus, and when sober senators accepted the dream as a reality and proceeded to legislate upon it, we can imagine the ideas that were passing in the minds of those of the general public who gave the subject a thought. Well, two years and more have now elapsed, and the unbounded potentialities of the new telegraphy have been whittled down by actual experiment to small practical though still very important proportions; and so, those interested in the old order can sleep in peace, and can go on doing so for a long time yet to come.

Having in the course of many years' researches in electric lore collected a mass of materials on this subject—for the idea embodied in the new telegraphy is by no means new—and having been a close observer of its recent and startling developments, I have thought that a popular account of its origin and progress would not now be uninteresting. This I have accordingly attempted in the following pages.

At an early stage in the evolution of our subject, objection was taken to the epithet Telegraphy without Wires, or, briefly, Wireless Telegraphy, as a misnomer (*e.g.*, the 'Builder,' March 17, 1855, p. 132), and in recent times the objection has been repeated. Induction, Space, and Ethereal Telegraphy have been suggested, but though accurate for certain forms, they are not comprehensive enough. A better name would be Telegraphy without Connecting Wires, which has also been suggested, but it is too cumbrous—an awkward mouthful. Pending the discovery of a better one, I have adhered to the original designation, Wireless Telegraphy, which actually is the popular one, and for which, moreover, I have the high sanction of her Majesty's Attorney-General.

In the course of a discussion on Mr (now Sir Wm.) Preece's paper on Electric Signalling without Wires ('Journal Society of Arts,' February 23, 1894), Sir Richard Webster laid down the law thus: "I think the objection to the title of the paper is rather hypercritical, because ordinary people always understand telegraphing by wire as meaning *through* the wire, going from one station to the other; and these parallel wires, not connected, would rather be looked upon as parts of the sending and receiving instruments. I hope, therefore, that the same name will be adhered to in any further development of the subject." If thus the name be allowable in Preece's case where, to bridge a space of, say, one mile, two parallel wires, each theoretically one mile long, are requisite, or double the amount required in the old form of telegraphy, it cannot be objected to in any of the other proposals which are described in these pages, certainly not to the Marconi system, where a few yards of wire at each end suffice for one mile of space, or, to put it accurately, where the height of the vertical wires (in yards) varies as the square root of the distance (in miles) to be signalled over.

At the outset of my task I was met with the difficulty of arranging my materials—whether in simple chronological order, or classified under heads, as Conduction, Induction, Wave, and Other or Miscellaneous Methods. Both have their advantages and disadvantages, but after consideration I decided to follow in the main the chronological order as the better of the two for a history which is intended to be a simple record of what has been done or attempted in the last sixty years by the many experimenters who have attacked the problem or contributed in any way to its solution.

Having settled this point, the further question of subdivision presented itself, and as the materials did not lend themselves to arrangement in chapters, I decided to divide the text into periods. The first I have called The Possible Period, which deals with first suggestions and empirical methods of experiment, and which, by reason of the want of delicacy in the instruments then available, may not inaccurately be compared with the Palæolithic period in geology. The second is The Practicable (or Neolithic) Period, when the conditions of the problem came to be better understood, and more delicate instruments of research were at hand. The third—The Practical Period—brings the subject up to date, and deals with the proposals of Preece (Electro-Magnetic), of Willoughby Smith (Conductive), and of Marconi (Hertzian), which are to-day in actual operation.

The whole concludes with five Appendices, containing much necessary information for which I could not conveniently find room in the body of the work. Appendix A deals with the philosophic views of the relation between electricity and light before and after Hertz, who, for the first time, showed them to be identical in kind, differing only in the degree of their wave-lengths. Appendix B gives in a popular form the modern views of electric currents consequent on the discoveries of Clerk-Maxwell, Hertz, and their disciples. Appendix C reproduces the greater part of Professor Branly's classic paper on his discovery of the Coherer principle, which is one of the foundation-stones of the Marconi system. Appendix D contains a very interesting correspondence between myself and Prof. Hughes, F.R.S., which came too late for insertion in the body of the work, and which is too important from the historical point of view to be omitted.

In Appendix E Mr Marconi's patent specification is reproduced, as, besides being historically interesting as the first patent for a telegraph of the Hertzian order, it is in itself a marvel of completeness. As the apparatus is there described, so it is used to-day after three years' rigorous experimentation, the only alterations being in points of detail—a finer adjustment of means to ends. This says much for the constructive genius of the young inventor, and bodes well for the survival of his system in the struggle for existence in which it is now engaged.

In the presentation of my materials I have allowed, as far as possible, the various authors to speak in their own words, merely condensing freely and, where necessary, translating obsolete words and phrases into modern technical language. This course in a historical work is, I think, preferable to obtruding myself as their interpreter. For the same reason I have given in the text, or in footnotes thereto, full references, so that the reader who desires to consult the original sources can readily do so.

I seem to hear the facetious critic exclaim, "Why, this is all scissors and paste." So it is, good sir, much of it; and so is all true history when you delete the fictions with which many historians embellish their facts. What one person said or what another did is not altered by the presence or absence of quotation marks. However, the only credit I claim is that due to collecting, condensing, and presenting my facts in a readable form—no light task,—and if my critics will award me this I will be satisfied.

Since the following pages were written, two excellent contributions have been made by Prof. Oliver Lodge and Mr Sydney Evershed in papers read before the Institution of Electrical Engineers, December 8 and 22, 1898. These will be found in No. 137 of the 'Journal,' and, together with the discussion which followed, should be studied by all interested in this fascinating subject. Mr Marconi has followed up these papers with one on his own method, which was read before the Institution on the 2nd of March last, and was repeated by general request on the

16th *idem.* He does not carry the matter farther than I have done in the text, but still the paper is worth reading —if only as an exposition in a nutshell of his beautiful system.

As a Frontispiece I give a group of twelve portraits of eminent men who may be fitly called the Arch-builders of Wireless Telegraphy. At the top stands Oersted (Denmark), who first showed the connection between electricity and magnetism. Then follow in order of time Ampère (France), Faraday (England), and Henry (America), who explained and extended the principles of the new science of electro-magnetism. Then come Clerk-Maxwell (England) and Hertz (Germany), who showed the relation between electricity and light, the one theoretically, and the other by actual demonstration. These are followed by Branly (France), Lodge (England), and Righi (Italy), whose discoveries have made possible the invention of Marconi. The last three are portraits of Preece and Willoughby Smith (England) and Marconi (Italy), who divide between them the honour of establishing the first practical lines of wireless telegraph—each typical of a different order.

ST HELIER'S, JERSEY,
September 1899.

CONTENTS.

FIRST PERIOD—THE POSSIBLE.

SECOND PERIOD—THE PRACTICABLE.

THIRD PERIOD—THE PRACTICAL.

SYSTEMS IN ACTUAL USE.

APPENDIX A.

APPENDIX B.

APPENDIX C.

APPENDIX D.

APPENDIX E.

A HISTORY OF
WIRELESS TELEGRAPHY.

FIRST PERIOD—THE POSSIBLE.

"Awhile forbear,

. . . .

Nor scorn man's efforts at a natural growth,
Which in some distant age may hope to find
Maturity, if not perfection."

PROFESSOR C. A. STEINHEIL—1838.

JUST mentioning *en passant* the sympathetic needle and sympathetic flesh telegraphs of the sixteenth and seventeenth centuries, a full account of which will be found in my 'History of Electric Telegraphy to 1837' (chap. i.),[1] we come to the year 1795 for the first glimmerings of telegraphy without wires. Salvá, who was an eminent Spanish physicist, and the inventor of the first electro-chemical telegraph, has the following bizarre passage in his paper "On the Application of Electricity to Telegraphy," read before the Academy of Sciences, Barcelona, December 16, 1795.

After showing how insulated wires may be laid under

[1] E. & F. N. Spon, London, 1884.

the seas, and the water used instead of return wires, he goes on to say: "If earthquakes be caused by electricity going from one point charged positively to another point charged negatively, as Bertolon has shown in his 'Électricité des Météores' (vol. i. p. 273), one does not even want a cable to send across the sea a signal arranged beforehand. One could, for example, arrange at Mallorca an area of earth charged with electricity, and at Alicante a similar space charged with the opposite electricity, with a wire going to, and dipping into, the sea. On leading another wire from the sea-shore to the electrified spot at Mallorca, the communication between the two charged surfaces would be complete, for the electric fluid would traverse the sea, which is an excellent conductor, and indicate by the spark the desired signal."[1]

Another early telegraph inventor and eminent physiologist, Sömmerring of Munich, has an experiment which, under more favourable conditions of observation, might easily have resulted in the suggestion at this early date of signalling through and by water alone. Dr Hamel[2] tells us that Sömmerring, on the 5th of June 1811, and at the suggestion of his friend, Baron Schilling, tried the action of his telegraph whilst the two conducting cords were each interrupted by water contained in wooden tubs. The signals appeared just as well as if no water had been interposed, but they ceased as soon as the water in the tubs was connected by a wire, the current then returning by this shorter way.

Now here we have, *in petto*, all the conditions necessary

[1] Later on (p. 81 *infra*) we shall see that Salvá's idea is after all not so extravagant as it seems. We now know that large spaces of the earth *can* be electrified, giving rise to the phenomenon of "bad earth," so well known to telegraph officials.

[2] 'Historical Account of the Introduction of the Galvanic and Electro-magnetic Telegraph into England,' Cooke's Reprint, p. 17.

for an experiment of the kind with which we are dealing, and had it been possible for Sömmerring to have employed a more delicate indicator than his water-decomposing apparatus he would probably have noticed that, notwithstanding the shorter way, some portion of the current still went the longer way ; and this fact could hardly have failed to suggest to his acute and observant mind further experiments, which, as I have just said, might easily have resulted in his recognition of the possibility of wireless telegraphy.

Leaving the curious suggestion of Salvá, which, though seriously meant, cannot be regarded as more than a *jeu d'esprit* — a happy inspiration of genius — and the what-might-have-come-of-it experiment of Sömmerring, we come to the year 1838, when the first intelligent suggestion of a wireless telegraph was made by Steinheil of Munich, one of the great pioneers of electric telegraphy on the Continent.

The possibility of signalling without wires was in a manner forced upon him. While he was engaged in establishing his beautiful system of telegraphy in Bavaria, Gauss, the celebrated German philosopher, and himself a telegraph inventor, suggested to him that the two rails of a railway might be utilised as telegraphic conductors. In July 1838 Steinheil tried the experiment on the Nürmberg-Fürth railway, but was unable to obtain an insulation of the rails sufficiently good for the current to reach from one station to the other. The great conductibility with which he found that the earth was endowed led him to presume that it would be possible to employ it instead of the return wire or wires hitherto used. The trials that he made in order to prove the accuracy of this conclusion were followed by complete success ; and he then introduced into electric telegraphy one of its greatest improvements—the earth circuit.[1]

[1] For the use of the earth circuit before Steinheil's *accidental* discovery, see my 'History of Electric Telegraphy,' pp. 343-345.

Steinheil then goes on to say: "The inquiry into the laws of dispersion, according to which the ground, whose mass is unlimited, is acted upon by the passage of the galvanic current, appeared to be a subject replete with interest. The galvanic excitation cannot be confined to the portions of earth situated between the two ends of the wire; on the contrary, it cannot but extend itself indefinitely, and it therefore only depends on the law that obtains in this excitation of the ground, and the distance of the exciting terminations of the wire, *whether it is necessary or not to have any metallic communication at all for carrying on telegraphic intercourse.*

"An apparatus can, it is true, be constructed in which the inductor, having no other metallic connection with the multiplier than the excitation transmitted through the ground, shall produce galvanic currents in that multiplier sufficient to cause a visible deflection of the bar. This is a hitherto unobserved fact, and may be classed amongst the most extraordinary phenomena that science has revealed to us. It only holds good, however, for small distances; and it must be left to the future to decide whether we shall ever succeed in telegraphing at great distances without any metallic communication at all. My experiments prove that such a thing is possible up to distances of 50 feet. For greater distances we can only conceive it feasible by augmenting the power of the galvanic induction, or by appropriate multipliers constructed for the purpose, or, in conclusion, by increasing the surface of contact presented by the ends of the multipliers. At all events, the phenomenon merits our best attention, and its influence will not perhaps be altogether overlooked in the theoretic views we may form with regard to galvanism itself." [1]

In another place, discussing the same subject, Steinheil

[1] Sturgeon's 'Annals of Electricity,' vol. iii. p. 450.

says: "We cannot conjure up gnomes at will to convey our thoughts through the earth. Nature has prevented this. The spreading of the galvanic effect is proportional, not to the distance of the point of excitation, but to the square of this distance; so that, at the distance of 50 feet, only exceedingly small effects can be produced by the most powerful electrical effect at the point of excitation. Had we means which could stand in the same relation to electricity that the eye stands to light, nothing would prevent our telegraphing through the earth without conducting wires; but it is not probable that we shall ever attain this end." [1]

Steinheil proposed another means of signalling without wires, which is curiously *àpropos* of Professor Graham Bell's photophone. In his classic paper on "Telegraphic Communication, especially by Means of Galvanism," he says: "Another possible method of bringing about transient movements at great distances, without any intervening artificial conductor, is furnished by radiant heat, when directed by means of condensing mirrors upon a thermo-electric pile. A galvanic current is called into play, which in its turn is employed to produce declinations of a magnetic needle. The difficulties attending the construction of such an instrument, though certainly considerable, are not in themselves insuperable. Such a telegraph, however, would only have this advantage over those [semaphores] based on optical principles—namely, that it does not require the constant attention of the observer; but, like the optical one, it would cease to act during cloudy weather, and hence partakes of the intrinsic defects of all semaphoric methods." [2]

[1] 'Die Anwendung des Electromagnetismus,' 1873, p. 172. We now have these means in "the electric eye" of Hertz! See pp. 180, 270 *infra*.

[2] 'Sturgeon's Annals of Electricity,' March 1839.

Acting on this suggestion, in June 1880 the present writer, while stationed at Teheran, Persia, and while yet ignorant of Professor Bell's method, worked out for himself a photophone, or rather a tele-photophone, which will be found described in the 'Electrician,' February 26, 1881. On my temporary return to England in 1882, I found that as early as 1878 Mr A. C. Brown, of the Eastern Telegraph Company, was working at the photophone. In September of that year he submitted his plans to Prof. Bell, who afterwards said of them: "To Mr Brown is undoubtedly due the honour of having distinctly and independently formulated the conception of using an undulatory beam of light, in contradistinction to a merely intermittent one, in connection with selenium and a telephone, and of having devised apparatus, though of a crude nature, for carrying it into execution" ('Jour. Inst. Elec. Engs.,' vol. ix. p. 404). Indeed the photophone is as much the invention of Mr Brown as of Prof. Bell, who, however, has all the credit for it in popular estimation.

EDWARD DAVY—1838.

While arranging, in 1883, the Edward Davy MSS., now in the library of the Institution of Electrical Engineers, the present writer discovered two passages which he at first took to have reference to some kind of telephonic relay; but on closer consideration it would appear that Davy had in view some contrivance based on the conjoint use of sound and electricity, much as Steinheil suggested the joint use of electricity and heat. The following are the passages to which I refer:—

At the end of a long critical examination of Cooke and Wheatstone's first patent of June 12, 1837, he says: "I have lately found that there is a peculiar way of propagating signals between the most distant places by self-acting means, and without the employment of any conducting wires at all. It is to be done partly by electricity, but combined with another principle, of the correctness of which there can be no doubt. But until I know what encouragement the other[1] will meet with I shall take no steps in this, as it may happen there will be other rivals. To give you a general idea of it, a bell may be rung at the first station, and then in the next instant a bell will ring at the next station a mile off, and so on for an unlimited series, though there is nothing between them but the plain earth and air! At the termination of the series, the signals may be given in letters, as in the present contrivance."

Again, in a paper of numbered miscellaneous memoranda, No. 20 reads as follows: "20. The plan proposed (101) of propagating communications by the conjoint agency of sound and electricity—the original sound producing vibrations which cause sympathetic vibrations in a unison-sounding apparatus at a distance, this last vibration causing a renewing wire to dip[2] and magnetise soft iron so as to repeat the sound, and so on in unlimited succession."

It is not easy to say from these passages (which are all we could find on the subject) what plan Davy had in contemplation. In the first quotation he speaks of bells, for which we may read a powerful trumpet at one end, and a concave reflector to focus the sound at the other

[1] That is, his chemical recording telegraph. See my 'History of Electric Telegraphy,' p. 379.

[2] *I.e.*, causing a relay to close a local circuit containing an electromagnet. Davy always spoke of the relay as the "renewer" or the "renewing wire"; and by dip he meant to dip into mercury, or, as we say nowadays, to close the circuit.

end; or some arrangement like the compressed-air telephone, proposed by Captain Taylor, R.N., in 1844; or the modern siren; or, in short, any means of producing sharp concussions of the air, such as were known in his day. Let us suppose he used any of these methods for projecting sound waves, then, at the focus of the distant reflector he may have designed a "renewing wire," so delicately poised as to respond to the vibration, and so close a local circuit in which was included the electro-magnetic apparatus for recording the sound, or for renewing it as required.

In the second passage he speaks of something on the principle of the tuning-fork. Now, tuning-forks in combination with reflectors may be practicable for short distances, but it is difficult to see how their vibrations could be utilised, at the distance of a mile, for "causing a renewing wire to dip."

However this may be, Davy's idea deserves at least this short notice in a history of early attempts at wireless telegraphy; for, although hardly possible of realisation with the apparatus at his command, it is perfectly feasible in these days of megaphones and microphones. As regards its practical utility, that is a question for the future, as to which we prefer not to prophesy.[1]

Davy's idea was probably the result of an incautious dose of the Auticatelephor of Edwards, which made a great stir a few years previously, and which, at first sight, might be taken to be a telegraph without *apparently* any

[1] Such a plan as Davy's was again suggested, in 1881, by Signor Senlicq d'Andres ('Telegraphic Journal,' vol. ix. p. 126), who, however, proposed to use, instead of a renewing wire or relay, the mouthpiece of a microphonic speaker, rendered more sensitive by a contact lever with unequal arms. Mr A. R. Sennett has also worked at the idea in more recent years. His method is very clearly described in the 'Jour. Inst. Elec. Engs.,' No. 137, p. 908.

connecting medium. We take the following announcement from the 'Kaleidoscope' of June 30, 1829 (p. 430):—

"THE AUTICATELEPHOR.

"We have received several papers descriptive of a new and curious engine, with the above name, invented by Mr T. W. C. Edwards, Lecturer on Experimental Philosophy and Chemistry, and designed for the instantaneous conveyance of intelligence to any distance. After noticing some of the greatest inventions of preceding times, Mr Edwards undertakes to demonstrate clearly and briefly, in the work which he has now in the press,[1] the practicability and facility of transmitting from London, *instantaneously*, to an agent at Edinburgh, Dublin, Paris, Vienna, St Petersburg, Constantinople, the Cape of Good Hope, Madras, Calcutta, &c., any question or message whatever, and of receiving back again at London, within the short space of one minute, an acknowledgment of the arrival of such question or message at the place intended, and a distinct answer to it in a few minutes. In principle this engine is altogether different from every kind of telegraph or semaphore, and requires neither intermediate station nor repetition. In its action it is totally unconnected with electricity, magnetism, galvanism, or any other subtle species of matter; and although the communication from place to place is instantaneous, and capable of ringing a bell, firing a gun, or hoisting a flag if required, yet this is not effected by the transit of anything whatever to and fro; nor in the operation is aught either audible or visible, except to the persons

[1] In 1883 we searched for this book in vain. Under the name T. W. C. Edwards we found in the British Museum Catalogue no less than twenty entries of translations from Greek authors, and of Greek and Latin grammars, &c.; but nothing to show that the writer was either a natural philosopher or a chemist.

communicating. It may be proper, however, to state that a channel or way must previously be prepared, by sinking a series of rods of a peculiar description in the ground, or dropping them in the sea; but these, after the first cost, will remain good for ages to come, if substantial when laid down."[1]

From the concluding words of this paragraph it would seem that the Auticatelephor was simply an application to telegraphy of pneumatic or hydraulic pressure in pipes—cautiously styled "rods of a peculiar description." On this supposition the last sentence may be paraphrased thus: "It may be proper, however, to state that a channel or way must previously be prepared, by laying down a continuous series of hollow rods or tubes under the ground or along the sea-bottom." If our supposition be correct, and if Edwards contemplated the use of compressed air, his proposal was certainly novel; but if he designed the use of compressed water, the idea was by no means new. Without going back to the old Roman plan of Æneas Tacticus, we have its revival by Brent and others towards the close of the last century, and the still more practical arrangements of Joseph Bramah in 1796, of Vallance in 1825, and of Jobard in 1827.

PROFESSOR MORSE—1842.

The idea of a wireless telegraph next appears to have presented itself to Professor Morse. In a letter to the Secretary of the Treasury, which was laid before the House of Representatives on December 23, 1844, he says:—

"In the autumn of 1842, at the request of the American

[1] See also the 'Mechanics' Magazine,' vol. xiii., First Series, p. 182.

Institute, I undertook to give to the public in New York a demonstration of the practicability of my telegraph, by connecting Governor's Island with Castle Garden, a distance of a mile; and for this purpose I laid my wires properly insulated beneath the water. I had scarcely begun to operate, and had received but two or three characters, when my intentions were frustrated by the accidental destruction of a part of my conductors by a vessel, which drew them up on her anchor, and cut them off. In the moments of mortification I immediately devised a plan for avoiding such an accident in future, by so arranging my wires along the banks of the river as to cause the water itself to conduct the electricity across. The experiment, however, was deferred till I arrived in Washington; and on December 16, 1842, I tested my arrangement across the canal, and with success. The simple fact was then ascertained that electricity could be made to cross a river without other conductors than the water itself; but it was not until the last autumn that I had the leisure to make a series of experiments to ascertain the law of its passage. The following diagram will serve to explain the experiment:—

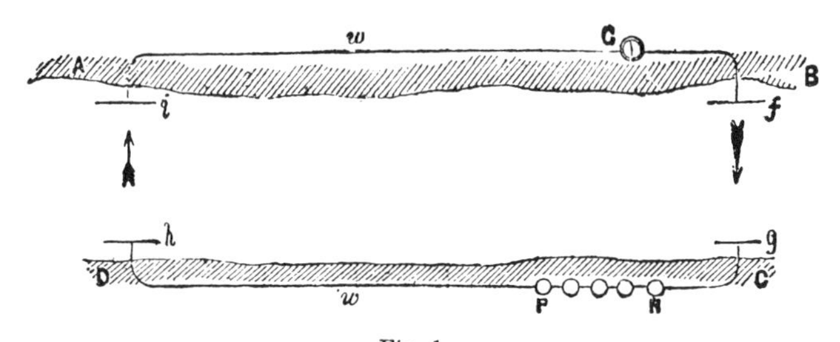

Fig. 1.

"A, B, C, D, are the banks of the river; N, P, is the battery; G is the galvanometer; *w w*, are the wires along the banks, connected with copper plates, *f*, *g*, *h*, *i*, which are placed in the water. When this arrangement is complete, the electricity, generated by the battery, passes from the positive pole, P, to the plate *h*, across the river through the water to plate *i*, and thence around the coil of the galvanometer to plate *f*, across the river again to plate *g*, and thence to the other pole of the battery, N.

"The distance across the canal is 80 feet; on August 24 the following were the results of the experiment:—

No. of the experiment.	1st.	2nd.	3rd.	4th.	5th.	6th.
No. of cups in battery .	14	14	14	7	7	7
Length of conductors, *w*, *w*	400	400	400	400	300	200
Degrees of motion of galvanometer . . .	32 & 24	13½ & 4½	1 & 1	24 & 13	29 & 21	21½ & 15
Size of the copper plates, *f*, *g*, *h*, *i*	5 by 2½ ft.	16 by 13 in.	6 by 5 in.	5 by 2½ ft.	5 by 2½ ft.	5 by 2½ ft.

"Showing that electricity crosses the river, and *in quantity in proportion to the size of the plates in the water.* The *distance of the plates on the same side* of the river *from each other* also affects the result. Having ascertained the general fact, I was desirous of discovering the best practical distance at which to place my copper plates, and not having the leisure myself, I requested my friend Professor Gale to make the experiments for me. I subjoin his letter and the results.[1]

"'NEW YORK, *Nov. 5th*, 1844.

"'MY DEAR SIR,—I send you herewith a copy of a series of results, obtained with four different-sized plates, as conductors to be used in crossing rivers. The batteries used were six cups of your smallest size, and one liquid

[1] We omit the tables of results, as of no present value. They can be seen in Vail's book, quoted *infra*.

used for the same throughout. I made several other series of experiments, but these I most rely on for uniformity and accuracy. You will see, from inspecting the table, that the distance along the shores should be *three times greater* than that from shore to shore across the stream; at least, that four times the distance does not give any increase of power. I intend to repeat all these experiments under more favourable circumstances, and will communicate to you the results.—Very respectfully, L. D. GALE.

"'Professor S. F. B. MORSE,
Superintendent of Telegraphs.'

"As the results of these experiments, it would seem that there may be situations in which the arrangements I have made for passing electricity across rivers may be useful, although experience alone can determine whether lofty spars, on which the wires may be suspended, erected in the rivers, may not be deemed the most practical. The experiments made were but for a short distance; in which, however, the principle was fully proved to be correct. It has been applied under the direction of my able assistants, Messrs Vail and Rogers, across the Susquehanna river, at Havre-de-Grace, with complete success, a distance of nearly a mile."[1]

JAMES BOWMAN LINDSAY—1843.

The next to pursue the subject was J. B. Lindsay of Dundee, whose extensive labours in this, as well as in the department of electric lighting, have hitherto been little appreciated by the scientific world. Through the kind assistance of Dr Robert Sinclair of Dundee, I have lately

[1] Vail's 'American Electro-Magnetic Telegraph,' Philadelphia, 1845.

collected a number of facts relating to this extraordinary man, and as I believe they will be new to most of my readers, I will draw largely from them in what follows.[1]

James Bowman Lindsay was born at Carmyllie, near Arbroath, on September 8, 1799, and but for the delicacy of his constitution would have been bred a farmer. At an early age he evinced a great taste for reading, and every moment that he could spare from his work as a linen-weaver was devoted to his favourite books. Often, indeed, he would be seen on his way to Arbroath with a web of cloth tied on his back and an open book in his hands; and, after delivering the cloth and obtaining fresh materials for weaving, he would return to Carmyllie in the same fashion. Encouraged by these studious habits, Lindsay's parents wisely arranged that he should go to St Andrews University. Accordingly, in 1821 he entered on his studies, and, self-taught though he had hitherto been, he soon made for himself a distinguished place among his fellow-students, particularly in the mathematical and physical sciences, in which departments, indeed, he became the first student of his time. Having completed the ordinary four years' course, Lindsay entered as a student of theology, and duly completed his studies in the Divinity Hall; but he never presented himself for a licence, his habits of thought inclining more to scientific than to theological pursuits. In the long summer vacations he generally returned to his occupation of weaving, though latterly he took up teaching, and thus enjoyed more time for the prosecution of his own studies.

Coming to Dundee in 1829, he was appointed Science and Mathematical Lecturer at the Watt Institution, then conducted by a Mr M'Intosh. Soon after, Alexander

[1] Extracts from the writer's articles in the 'Electrical Engineer,' vol. xxiii. pp. 21, 51.

Maxwell, the historian of Dundee, became a pupil, and this is the picture he has left us of Lindsay :—

"When I was with Mr M'Intosh, I attended classes that were taught by Mr Lindsay, a man of profound learning and untiring scientific research, who, had he been more practical, less diffident, and possessed of greater worldly wisdom, would have gained for himself a good place amongst distinguished men. As it was, he remained little more than a mere abstraction, a cyclopædia out of order, and went through life a poor and modest schoolmaster.

"By the time I knew him he was devoting much of his attention to electricity, to the celerity with which it was transmitted to any distance, and to the readiness with which its alternating effects may be translated into speech —and I have no doubt he held in his hand the modern system of the telegraph, but it needed a wiser man than he to turn it to practical use. He also produced from galvanic cells a light which burned steadily for a lengthened period.

"His acquaintance with languages was extraordinary, and almost equalled that of his famous contemporary, the Cardinal Mezzofanti. In 1828 he began the compilation of a dictionary in fifty languages, the object of which was to discover, if possible, by language the place where, and the time when, man originated. This stupendous undertaking, which occupied the main part of his life's work, he left behind in a vast mass of undigested manuscript, consisting of dissertations on language and cogitations on social science—a monument of unpractical and inconclusive industry. In 1845 he published 'A Pentecontaglossal Paternoster,' intended to serve as a specimen of his fifty-tongued lexicon.

"In 1858 he published 'The Chrono-Astrolabe,' for determining with certainty ancient chronology—a work on which he had been engaged for many years; and in 1861

'A Treatise on Baptism,' which is a curious record of his philosophical knowledge. . . .

"In 1832 he obtained a situation as travelling tutor, which was to take him abroad for some time. We loved him as much as consists with a boy's nature to love his teacher, and subscribed for a silver snuff-box as a slight mark of our regard. . . .

"I am afraid that the situation of travelling tutor did not turn out well, for within two years Lindsay was back again in Dundee, and resumed his position of assistant teacher, arduously following at the same time his favourite studies."

The scope of his teaching at this time is shown by the following notice which appeared in the 'Dundee Advertiser' of April 11, 1834:—

"J. B. Lindsay resumes classes for cultivating the intellectual and historical portions of knowledge and instruction on April 14, 1834, in South Tay Street, Dundee.

"In a few weeks hence a course of lectures will be formed on frictional, galvanic, and voltaic electricity; magnetism; and electro-magnetism. The battery, already powerful, is undergoing daily augmentation. The light obtained from it is intensely bright, and the number of lights may be increased without limit.

"A great number of wheels may be turned [by electricity], and small weights raised over pulleys.

"Houses and towns will in a short time be lighted by electricity instead of gas, and heated by it instead of coal; and machinery will be worked by it instead of steam—all at a trifling expense.

"A miniature view of all these effects will be exhibited, besides a number of subordinate experiments, including the discoveries of Sir Humphry Davy."

In March 1841, Lindsay was appointed teacher in the

Dundee Prison on a salary of £50 a-year, a post which he held for upwards of seventeen years, till October 1858. It is stated that shortly after taking up this office he could have obtained an appointment in the British Museum, a situation which would have been most congenial to his tastes, and which would certainly have led to a lasting recognition of his great abilities; but, being unwilling to leave his aged mother, he declined the offer—a rare example of devotion and self-denial. . . .

Lindsay was a bachelor, and lived alone, buried, it might be said, in his books, collections of which, in history and philosophy, science and languages, were heaped in every corner of his dwelling—a small house of three apartments (11 South Union Street). The kitchen was filled with electrical apparatus, mostly the work of his own hands; and his little parlour was so crowded with books, philosophical apparatus, and other instruments of his labour, that it was difficult to move in it. To provide these things, he denied himself through life the ordinary comforts and conveniences,—bread and coffee, and other simple articles, forming the principal part of his diet. His house in time acquired a celebrity as one of the curiosities of Dundee, and men of learning from distant parts, not only of the kingdom but of the world, often came to pay him a visit.

. . . . . .

In July 1858, on the recommendation of Lord Derby, then Prime Minister, her Majesty granted Lindsay an annual pension of £100 a-year, "in recognition of his great learning and extraordinary attainments." This well-deserved bounty relieved him from the drudgery of a prison teacher, and henceforth to the close of his life he devoted himself entirely to literary and scientific pursuits.

Although never robust, Lindsay on the whole enjoyed tolerably good health through life, but trouble came at last.

On June 24, 1862, he was seized with diarrhœa, which carried him off on June 29, 1862, in the sixty-third year of his age.[1]

Although languages and chronology took up much (I am inclined to think too much) of Lindsay's time, still electricity and its applications were his first, as they were always his favourite, study. Amongst some notes and memoranda, bound up with his manuscripts in the Albert Institute, Dundee, he says:—

"Previous to the discovery of Oersted, I had made many experiments on magnetism, with the view of obtaining from it a motive power. No sooner, however, was I aware of the deflection of the needle and the multiplication of the power by coils of wire than the possibility of power appeared certain, and I commenced a series of experiments in 1832. The power on a small scale was easily obtained, and during these experiments I had a clear view of the application of electricity to telegraphic communication. The light also drew my attention, and I was in a trilemma whether to fix upon the power, the light, or the telegraph. After reflection I fixed upon the light as the first investigation, and had many contrivances for augmenting it and rendering it constant. Several years were spent in experiments, and I obtained a constant stream of light on July 25, 1835. Having satisfied myself on this subject, I returned to some glossological investigations that had been left unfinished, and was engaged with these till 1843. In that year I proproposed a submarine telegraph across the Atlantic, after having proved the possibility by a series of experiments. Inquiries on other subjects have since that time engaged my attention, but I eagerly desire to return to electricity."

The first public announcement of Lindsay's success in

[1] Norrie's 'Dundee Celebrities of the Nineteenth Century,' Dundee 1873.

electric lighting was contained in a short paragraph in the 'Dundee Advertiser' of July 31, 1835; and on October 30 following the same paper published a letter on the subject from Lindsay himself:—

"ELECTRIC LIGHT.

"SIR,—As a notice of my electric light has been extensively circulated, some persons may be anxious to know its present state, and my views respecting it.

"The apparatus that I have at present is merely a small model. It has already cost a great deal of labour, and will yet cost a good deal more before my room is sufficiently lighted. Had circumstances permitted, it would have been perfected two years ago, as my plans were formed then. I am writing this letter by means of it, at 6 inches or 8 inches distant; and, at the present moment, can read a book at the distance of 1½ foot. From the same apparatus I can get two or three lights, each of which is fit for reading with. I can make it burn in the open air, or in a glass tube without air, and neither wind nor water is capable of extinguishing it. It does not inflame paper nor any other combustible. These are facts.

"As I intend in a short time to give a lecture on the subject, my views on the further progress will be unfolded then. A few of these, however, may be mentioned just now.

"Brilliant illumination will be obtained by a light incapable of combustion; and, on its introduction to spinning mills, conflagrations there will be unheard of. Its beauty will recommend it to the fashionable; and the producing apparatus, framed, may stand side by side with the piano in the drawing-room. Requiring no air for combustion, and emitting no offensive smell, it will not deteriorate the atmosphere in the thronged hall. Exposed to the open day, it will blaze with undiminished lustre amidst tempests of wind and rain; and, being capable of surpassing all lights in splendour, it will be used in lighthouses and for telegraphs. The present generation may yet have it burning in their houses and enlightening their streets. Nor are these predictions the offshoots of an exuberant fancy or disordered imagination. They are the anticipated results of laborious research and of countless experiments. Electricity, moreover, is destined for mightier feats than even universal illumination. J. B. LINDSAY.

"DUNDEE, *Oct.* 28, 1835."

Lindsay's connection with electric telegraphy forms a very interesting episode. We have seen that from about the year 1830 he was familiar with telegraphic projects, and that he made them the subject of illustration in his classes. At this date electric telegraphs were distinctly in the air, but, like electric lighting, they had hardly advanced beyond the laboratory stage.[1] Lindsay does not appear to have carried them much further for several years, for it was not until 1843 that he conceived the bold idea of a submarine telegraph to America by means of a naked wire and earth-batteries, "after having proved the possibility by a series of experiments."

It is true that at this time the earth-battery was known. It was first proposed by Kemp, of Edinburgh, in 1828; Prof. Gauss in 1838 suggested its employment for telegraphic purposes, and Steinheil, acting on the suggestion, actually used it with some success on the Munich-Nanhofen

[1] From the public exhibition of Baron Schilling's needle instrument in Germany in 1835-36 dates the first real start of electric telegraphy. See my 'History of Electric Telegraphy,' chap. ix.

Railway, twenty-two miles long; and Bain in October 1842 employed it for working clocks. Similarly, the idea of signalling with uninsulated wire and without any wire at all was not new, for, as we have seen, the possibility of doing so was in a manner forced on the notice of Steinheil in 1838 and on Morse in 1842, but Lindsay was certainly the first to combine the two principles in his daring proposal of an Atlantic telegraph; and this, be it remembered, at a time when electric telegraphy was still a young and struggling industry, and when submarine telegraphy was yet a dream.

On June 19, 1845, a short paragraph appeared in the 'Northern Warder,' Dundee, referring to a New York project of communicating between England and America by means of a submerged copper wire "properly covered and of sufficient size." This called forth the following letter from Lindsay, which was published in the same paper on June 26 following:—

"ELECTRIC TELEGRAPH TO AMERICA.

"SIR,—The few lines I now send you have been occasioned by a notice in your last in reference to an electric telegraph to America. Should the plan be carried into effect the following hints should be attended to: The wire should be of pure copper, as otherwise it would be injured by the electro-chemical action of the water. The wire must not be composed of parts joined by soldering, but welded together; this welding can be performed by electricity. In order to prevent the action of water on the wire, a button of a more oxidable metal should be welded to it at short distances; the best metal for this purpose would be lead. If soldered to the wire, it must be soldered by lead alone. No third metal must be used. If welded, it may be done by electricity. In this way the wire resting on the bottom of the sea might last a long time. The one end of the wire is then to be soldered or welded to a plate of zinc immersed in the ocean on the coast of Britain, and the other end similarly joined to a plate of copper deposited in the same ocean on the coast of America. In reference to the expense, suppose the wire to be a ninth or tenth of an inch diameter, then the length of 100 inches would contain a cubic inch of copper, and three miles of wire would contain a cubic foot, weighing 9000 ounces, of the value of about £36 sterling. Owing to the inequalities in the bottom of the ocean, the distance to America might be 3000 miles, and the expense £36,000 sterling—a trifle when compared with the resulting benefit. The only injury that the wire is likely to undergo is from submarine eruptions. It may be broken by these. The two ends, however, being accessible, the greater part of the wire may be drawn up, and the necessary length of wire welded to it. It should be remembered that this welding must be done by electricity. To Calcutta, by the Cape of Good Hope, the expense would be £200,000. The wire from Calcutta to Canton would cost £70,000, to New Zealand £120,000, to Tahiti nearly £200,000. A wire might be placed round the coast of Britain, and another along the coast of America. There might be stations at different towns and electric clocks agreeing with each other to a second of time. Each town might have a specific time for intelligence. Suppose Dundee to have the hour from nine to ten. From nine to ten minutes past nine, messages are sent and answers received between Dundee and New York. From ten minutes to twenty minutes past nine communication is made between Dundee and Quebec. The rest of the hour is for intercourse between Dundee and other towns. The same is done with Edinburgh,

Glasgow, Liverpool, &c., each town having an hour for itself.—L.

"DUNDEE, *June* 21, 1845."

From this letter it is clear that Lindsay then contemplated an uninsulated wire across the Atlantic in connection with what have come to be known as earth-batteries at the stations along the coasts. His plan of protecting the wire from the corrosive action of the sea-water was evidently borrowed from Sir Humphry Davy's proposal of 1824 for the protection of the copper sheathing of ships by strips of zinc; while the further suggestion, on which he insists so much, of welding the various lengths of wire by electricity, if not original with him, was at all events a very early recognition of a process which has cropped up again in recent years, and which is now largely employed.[1]

Between 1845 and 1853 Lindsay does not appear to have done anything in furtherance of his Atlantic project, being probably wholly absorbed in his linguistic and chronological studies. At all events, we hear nothing from him until March 11, 1853, when a notice appeared in the 'Dundee Advertiser' of a lecture which he proposed to give on the ensuing Tuesday at the Thistle Hall.

In the same paper a week later a report of the lecture is given as follows :—

"TELEGRAPHIC COMMUNICATION.

"On Tuesday evening our learned and ingenious townsman, Mr J. B. Lindsay, delivered a lecture on the above subject, one with which he has an acquaintance second to

[1] Electric welding was proposed by Joule in 1856; by Wilde in 1865; and by Prof. Elihu Thomson (America) and Dr Benardos (Russia) in 1887.

no man in the kingdom. It would be impossible, in the limited space at our disposal, to give any *vidimus* of the lecture; we can only indicate the outline of a recent discovery made by Mr Lindsay, involving a principle which, if capable of acting irrespective of distance (and we see no reason to doubt that it is), must by-and-by revolutionise all our ideas of time and space. Mr Lindsay stated the principle to be that submerged wires, such as those now used for telegraphic intelligence between this country and Ireland and France, were no longer necessary. By a peculiar arrangement of the wires at the sides of rivers or seas, the electric influence can be made to pass on through the water itself. This proposition was certainly startling, but he illustrated it on a small scale by means of a water-trough, and, so far as the experiment went, it faithfully developed the principle. Mr Lindsay, after concluding these experiments, proceeded to point out the lines which appeared to him most eligible for transmitting telegraphic intelligence throughout the world; and, having done so, he wound up with a peroration of great beauty, in which the wonders to be achieved by electric influence in the days to come were eloquently set forth. It is a fine sight to see this learned and philosophic man pursuing the studies of science and literature, not for the sake of any empty applause, but for those pure pleasures they are in themselves so well fitted to bestow. At the same time, it is gratifying to know that there are many people capable of appreciating the modest and retiring character of Mr Lindsay,—a fact which was clearly evidenced on Tuesday evening by the numerous and most respectable meeting which then assembled to hear his scientific lecture."

In the following August Lindsay delivered another lecture (probably the same) in Glasgow, and so sanguine was he at this time of the practicability of his method that he actually patented it on June 5, 1854. The following account, which I have condensed from the specification of his patent, explains the *modus operandi*, and also shows how well he understood the conditions of the problem :—

"My invention consists of a mode of transmitting telegraphic messages by means of electricity or magnetism through and across water without submerged wires, the water being made available as the connecting and conducting medium by the following means :—

"On the land, on the side from which the message is to be sent, I place a battery and telegraph instrument, to which are attached two wires terminating in metal balls, tubes, or plates placed in the water or in moist ground adjacent to the water at a certain distance apart, according to the width of the water to be crossed (the distance between the two balls, plates, or tubes to be greater than across the water when practicable). On the land which is situated on the opposite side of the water, and to which the message is to be conveyed, I place two similar metal balls, plates, or tubes, immersed as above stated, and having wires attached to them which lead to, and are in connection with, another battery and needle indicator, or other suitable telegraphic instrument. A, A in the diagram (fig. 2) show the position of the battery and instrument on one side of the water, Z; B, B, the battery and instrument on the opposite side; C, D, E, F, metallic or charcoal terminators; G, H, I, K, wires insulated in the usual way, and connecting the terminators, batteries, and instruments, as shown.

"As regards the power or primary agent, it may be either voltaic, galvanic, or magnetic electricity, and the apparatus for evolving the same, such as is used for ordinary telegraphic purposes.

"As regards the indicating apparatus, I propose to employ any of the instruments in known use which are most efficient for my purpose, observing that the needle indicator may be arranged either in a vertical or in a horizontal position, and that the coil of wire which actuates the needle may be increased or diminished according to circumstances.

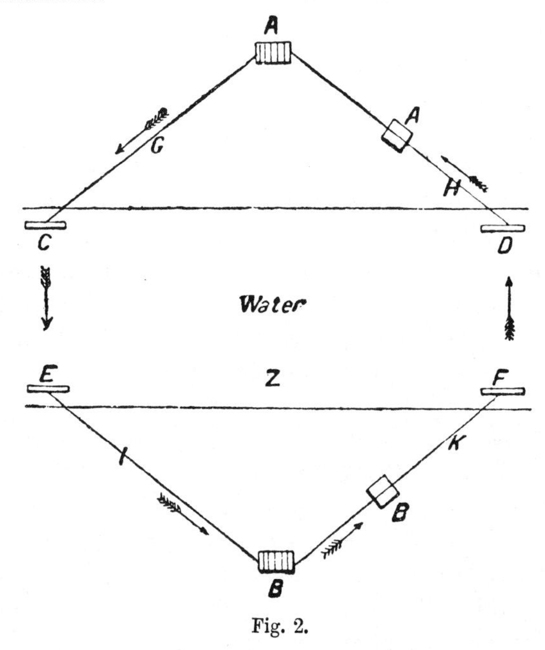

Fig. 2.

"Suppose it is required to transmit a message from A, the operator completes the circuit of the electric current as ordinarily practised. It will be evident that the current will have two courses open to it, the one being directly back through the water from C to D, and the other across the water from C to E, along the wires I K, through the instrument B, and back from F to D. Now, I have found that if each of the two distances C D and E F be greater than C E and D F, the resistances through C E and D F will be so much less than that through the water between C and D, that more of the current will pass across the water, through the opposite wires, and recross at F, than take the direct course C D; or, more correctly speaking, the current will divide itself between the two courses in inverse ratio to their resistances. As cases may arise, from local or other causes, such as not to admit of the distance between the immersed plates being greater than the distance across the water, I propose, then, to augment the force of the batteries, and to increase the size of the plates, so as to compel a sufficient portion of the current to cross. I prefer, however, when circumstances admit of it, employing the first method."

Lindsay's first public trials were across the Earl Grey Docks at Dundee, and then across the Tay at Glencarse, where the river is nearly three-quarters of a mile wide. Of the few friends who assisted at these experiments Mr Loudon of Dundee is, I believe, the only one now left. He tells us that Lindsay would station them on one side of the Tay, enjoining them to watch the galvanometer and note down how the needle moved. He would then insert his plates in the water on their side of the river, and, crossing over to the opposite side, would complete his arrangements. With a battery of twenty-four Bunsen cells he would make a few momentary contacts, reversing the connections a few times so as to produce right and left deflections of the galvanometer needle. Then he would return and compare the deflections of the needle which they had noted with the order in which he had himself made the battery contacts, and on finding them to correspond he would be supremely happy.[1]

In 1854 Lindsay was in London, and brought his plans to the notice of the Electric Telegraph Company. It is

[1] Kerr, 'Wireless Telegraphy,' 1898, p. 40.

now curious to remark that Sir W. H. Preece, who, as we shall see later on, became himself in after years an eminent wireless-telegraph inventor, was the officer who was deputed to assist him and report on his method. Sir William tells us that these were almost the first electrical experiments of any importance in which he ever took part, and in a letter to the writer, dated October 15, 1898, he adds: "I remember Lindsay very well. He came up to London with his 'great invention,' and I assisted him in making his experiments in our gutta-percha testing tank at Percy Wharf on the Thames. We used the old sand battery and galvanometers—ohms and volts were not invented then—and showed that by varying the distance apart of the plates on each side of the tank we varied the strength of the signals. I have no record of the results, but they showed the feasibility of the plan. I had, however, to crush poor Lindsay by telling him that it was not new. Morse in 1842 had done the same thing, and Alexander Bain had also tried about the same time a similar experiment on the Serpentine, but I have not found any published record of it." [1]

In August 1854 Lindsay carried out a series of experiments at Portsmouth, in which, according to a notice in the 'Morning Post' (August 28), he completely succeeded in transmitting signals across the mill dam, where it is about 500 yards wide.[2]

[1] In this, I think, his memory betrays him. Bain's experiments had to do with an insulated wire in connection with earth-batteries. See 'The Artisan,' June 30, 1843, p. 147.

[2] These experiments were also noticed in 'Chambers's Journal' for September 1854, as follows: "Again has an attempt been made to send a signal through water without a wire—this time at Portsmouth, where it was attended with partial success. The thing has often been tried: a few years ago, a couple of *savants* might have been seen sending their messages across those minor lakes known to

Lindsay repeated these experiments at intervals and at various places, indeed whenever and wherever he had the chance, his greatest performance being across the Tay, from Dundee to Woodhaven, where the river is nearly two miles broad. On one of these occasions, and when an Atlantic telegraph began to be seriously debated, the difficulty of finding a steamer large enough to carry the cable was discussed, when Lindsay quietly remarked, "If it were possible to provide stations at not more than twenty miles distant all the way across the Atlantic, I would save them the trouble of laying any cable."

In September 1859 Lindsay read a paper before the British Association at Aberdeen "On Telegraphing without Wires," which drew from Lord Rosse, the president of the section, special commendation. Prof. Faraday and (Sir) G. B. Airy, then Astronomer-Royal, also added their approval of the views enunciated. Prof. Thomson (now Lord Kelvin) was also present, and, as is well known, was then deeply engaged with Atlantic cable projects. History does not say what *he* thought of the poor Dundee lecturer, but, with the experience of forty years, we can easily guess.

A brief abstract of the paper was published in the Annual Report of the Association for 1859, but a fuller account appeared in the 'Dundee Advertiser,' from which I take the following interesting details:—

"The author has been engaged in experimenting on the

Londoners as the Hampstead Ponds!" Can any reader tell me who these *savants* were?

About this time experiments in wireless telegraphy were evidently popular. Van Reese at Portsmouth; Gintl, the first inventor of a duplex telegraph, in Austria; Bonelli in Italy, and Bouchotte and Douat in France (and doubtless others), all were engaged on the problem, but with what results I do not know, as I have not met with any detailed accounts of their experiments.

subject, and in lecturing on it in Dundee, Glasgow, and other places since 1831. Recently he had made additional experiments, and succeeded in crossing the Tay where it was three-quarters of a mile broad. His method had always been to immerse two plates or sheets of metal on the one side, and connect them by a wire passing through a coil to move a needle, and to have on the other side two sheets similarly connected, and nearly opposite the two former. Experiments had shown that only a fractional part of the electricity generated goes across, and that the quantity that thus goes across can be increased in four ways: (1) by an increased battery power; (2) by increasing the surface of the immersed sheets; (3) by increasing the coil that moves the receiving needle; and (4) by increasing the lateral distance of the sheets. In cases where lateral distance could be got he recommended increasing it, as then a smaller battery power would suffice. In telegraphing by this method to Ireland or France abundance of lateral distance could be got, but for America the lateral distance in Britain was much less than the distance across. In the greater part of his experiments the distance at the sides had been double the distance across; but in those on the Tay the lateral distance was the smaller, being only half a mile, while the distance across was three-quarters of a mile.

"Of the four elements above mentioned, he thought that if any one were doubled the portion of electricity that crossed would also be doubled, and if all the elements were doubled the quantity transmitted would be eight times as great. In the experiments across the Tay the battery was of 4 square feet of zinc, the immersed sheets contained about 90 square feet of metal, the weight of the copper coil was about 6 lb., and the lateral distance was, as just stated, less than the transverse; but if it had been a mile, and the distance across also a mile, the signals would, no doubt, have been equally distinct. Should this law (when the lateral distance is equal to the transverse) be found correct, the following table might then be formed:—

Zinc for battery.	Immersed sheets.	Weight of coil.	Distance crossed.
sq. ft.	sq. ft.	lb.	miles.
4	90	6	1
8	180	12	8
16	360	24	64
32	720	48	512
64	1440	96	4096
128	2880	192	32,768

"But supposing the lateral distance to be only half the transverse, then the space crossed might be 16,000 miles; and if it was only a fourth, then there would be 8000 miles—a much greater distance than the breadth of the Atlantic. Further experiments were, however, necessary to determine this law, but, according to his calculations, he thought that a battery of 130 square feet, immersed sheets of 3000 square feet, and a coil of 200 lb.,[1] would be sufficient to cross the Atlantic with the lateral distance that could be obtained in Great Britain."

After the reading of the paper Lindsay carried out some very successful experiments across the river Dee, in the

[1] My readers will smile at the suggestion of such galvanometer coils, but they should remember that forty years ago matters electrical were largely ordered by the rule of thumb. The electro-magnet first used by Morse on the Washington-Baltimore line (1844), and exhibited in Europe, weighed 185 lb. The arms were 3½ inches long and 18 inches diameter, the wire (copper) being that known as No. 16—the same size as the line wire, it being then supposed that the wire of the coils and of the line should be of the same size throughout. Down to 1860 not a few practical telegraphists held this view. See D. G. FitzGerald in the London 'Electrical Review,' August 9, 1895, p. 157.

presence of Lord Rosse, Prof. Jacobi of St Petersburg, and other members of the Association. In February 1860 he made Liverpool the scene of his operations, but there, strange to say, he had not the success which hitherto attended him. The experiments failed, being "counteracted by some unaccountable influence which he had not before met with." However, in the following July he was again successful at Dundee in his experiments across the Tay, below the Earn, where the river is more than a mile wide. In communicating these results to the 'Dundee Advertiser' (July 10, 1860), he says: "The experiment was successful, and the needle was strongly moved; but as I had no person with me capable of sending or reading a message, it [regular telegraphic signalling] was not attempted."

This was Lindsay's last public connection with the telegraph, but to the end of his life (June 29, 1862) he remained perfectly convinced of the soundness of his views and of their ultimate success.[1]

[1] On the eve of the centenary of Lindsay's birth the 'Dundee Advertiser' (September 7, 1899) published a very appreciative sketch of "the famous Scottish inventor," which is largely based on my articles quoted on p. 14, *supra*. As a result we are gratified to learn that "a bust of James Bowman Lindsay, a pioneer of wireless telegraphy by the conductive method, is to be placed in the Victoria Art Galleries of Dundee. The bust is to be of white Carrara marble, and will be the gift of Lord Provost M'Grady, Mr George Webster of Edinburgh being the sculptor. It has further been proposed to erect a monument over Lindsay's grave by public subscription."—'Electrician,' vol. xliii. p. 795.

J. W. WILKINS—1845.

In the New York 'Electrical Engineer' of May 29, 1895, it was claimed for Prof. Trowbridge (of whom we shall have more to say later on) that he was the first to telegraph without wires in 1880.

The paragraph in which this claim, unfounded as we already see, was advanced, besides drawing renewed attention to Prof. Trowbridge's experiments, had the merit of calling forth an interesting communication from our own Mr J. W. Wilkins, one of the very few telegraph officers of Cooke & Wheatstone's days still with us, and whose early and interesting reminiscences I hope we may yet see.[1]

Writing in 'The Electrician,' July 19, 1895, Mr Wilkins says:—

"Nearly fifty years ago, and thirty years before Prof. Trowbridge 'made original researches between the Observatory at Cambridge and the City of Boston,' the writer of these lines had also researched on the same subject, and a year or two later published the results of his investigations in an English periodical—the 'Mining Journal' of March 31, 1849—under the heading 'Telegraph communication between England and France.' In that letter, after going into the subject very much like the American Professor in 1880, there will be found my explanation—also not differing much from the Professor's—as to how the thing was to be done; except that, in my case, I proposed a new and delicate form of galvanometer or telegraph instrument for the purpose, while he made use of the well-known telephone. I suggested the erection of lengths of telegraph wires on the

[1] Mr Wilkins is the author of two English patents: (1) Improvements in Electric Telegraphs, January 13, 1853; and (2) Improvements in obtaining power by Electro-Magnetism, October 28, 1853.

English and French coasts, with terminals dipping into the earth or sea, and as nearly parallel as possible to one another; and I suggested a form of telegraph instrument consisting of 'coils of finest wire, of best conductibility,' with magnets to deflect them on the passage of a current of electricity through them, which I expected would take place on the discharge of electricity through the circuits on either side of the water; anticipating, of course, that a portion of the current would flow from the one pair of earth-plates—terminals of one circuit—to the other pair of terminals on the opposite shore.

. . . . . .

"It may be interesting to relate how I came to think that telegraphy without wires was a possibility, and that it should have appeared to me to have some value, at a time when gutta-percha as an insulator was not imagined, or the ghost of a proposition for a submarine wire existed. At that time, too, it was with the utmost difficulty that efficient insulation could be maintained in elevated wires if they happened to be subject to a damp atmosphere.

"It was in the year 1845, and while engaged on the only long line of telegraph then existing in England—London to Gosport — that my observations led me to question the accepted theory that currents of electricity, discharged into the earth at each end of a line of telegraph, sped in a direct course—instinctively, so to say—through the intervening mass of ground to meet a current or find a corresponding earth-plate at the other end of it to *complete the circuit.* I could only bring myself to think that the earth acted as a reservoir or condenser—in fact, receiving and distributing electricity almost superficially for some certain or uncertain distance around the terminal earths, and that according to circumstances only. A year later, while occupied with the installation of telegraphs for Messrs Cooke & Wheatstone (afterwards the Electric Telegraph Company), a good opportunity offered of testing this matter practically upon lengths of wire erected on both sides of a railway. To succeed in my experiment, and detect the very small amount of electricity likely to be available in such a case, I evidently required the aid of a very sensitive galvanometer, much more so indeed than the long pair of astatic needles and coil of the Cooke & Wheatstone telegraph, which was then in universal use as a detector. The influence of magnetism upon a wire conveying an electric current at once suggested itself to me, and I constructed a most sensitive instrument on this principle, by which I succeeded in obtaining actual signals between lengths of elevated wires about 120 ft. apart. This, however, suggested nothing more at the moment than that the current discharged from the earth-plates of one line found its way into the earth-plates of another and adjacent circuit, through the earth. Later on, I had other opportunities of verifying this matter with greater distances between the lines of wire, and ultimately an instance in which the wires were a *considerable distance apart*, and with no very near approach to parallelism in their situation. Then it was that it entered my head that telegraphing without wires *might be a possibility.*"

. . . . . .

The following extracts from the letter in the 'Mining Journal,' above referred to, may now be reproduced with interest. I have slightly altered the phraseology with a view of making the writer's meaning more clear and connected :[1]—

"Allow me, through the medium of your valuable

[1] Mr Charles Bright has reprinted this letter *verbatim* in Jour. Inst. Elec. Engs.,' vol. xxvii. p. 958, as containing "the first really practical suggestion in the direction of inductive telegraphy"; but, as we now see, it is not the first suggestion, and it is certainly not inductive.

journal, to draw attention to a principle upon which a telegraphic communication may be made between England and France without wires. I take for certain (as experiments I have made have shown me) that when the poles of a battery are connected with any extended conducting medium, the electricity diffuses itself in radial lines between the poles. The first and larger portion will pass in a straight line, as offering the least resistance; the rays will then form a series of curves, growing larger and larger, until, by reason of increasing distance, the electricity following the outer curves is so infinitesimal as to be no longer perceptible.

"These rays of electricity may be collected within a certain distance—focussed as it were—by the interposition of a metallic medium that shall offer less resistance than the water or earth; and, obviously, the nearer the battery, the greater the possibility of collecting them. I do not apprehend the distance of twenty miles being at all too much to collect a sufficient quantity of electricity to be useful for telegraphic purposes. If, then, it is possible, as I believe, to collect in France some portion of the electricity which has been discharged from a battery in England, all that is required is to know how to deal with it so that it shall indicate its presence.

"The most delicate of the present telegraph apparatus, the detector, being entirely unsuited for the purpose, I propose the following arrangement: Upon one shore I propose to have a battery that shall discharge its electricity into the earth or sea, with a distance between its poles of five, ten, or twenty miles, as the case may be. Let a similar length of wire be erected on the opposite coast, as near to, and parallel with, it as possible, with its ends also dipping into the earth or sea. In this circuit place an instrument consisting of ten, twenty, or more round or square coils of the finest wire of best conductibility, suspended on points or otherwise between, or in front of, the poles of an electro-, or permanent, magnet or magnets. Any current passing through the coil would be indicated by its moving or shifting its position with reference to the poles of the magnet. This would constitute a receiving apparatus of the most delicate character, for its efficiency would depend not so much on the strength of the current passing as on the power of the magnet, which may be increased at pleasure.

"I hope some one will take up this suggestion and carry it out practically to a greater extent than my limited experiments have enabled me to do. Of its truth for long as well as for short distances I am satisfied, and only want of means and opportunity prevent me carrying it out myself."

In a recent letter to the writer *àpropos* of this early proposal, Mr Wilkins says:—

"I will just say that all thought of induction was absent in my first experiments. I modified my views in this respect a year or two later, but I did not attach sufficient importance to the matter to follow up my communication to the 'Mining Journal,' especially as at that time a cable was actually laid across the Channel, which I could not doubt would be a success, and a permanent one too. I rather courted forgetfulness of the proposition. Whatever my opinion at the time was as to the source of the electricity that I discovered in the far removed and disconnected circuit, the result was the same, and the means I used to obtain it the same in principle as those which make the matter an accomplished fact to-day—viz., elevated lengths of wire, and the discharge of electricity from the one on to a delicate receiving apparatus in the circuit of the other.

"As regards the form of receiving apparatus which I suggested for indicating the signals, I did then, and do now, attach great importance to the happy idea. It happens to be the most delicate form of detector or galvanometer, and is identical in principle with Lord Kelvin's apparatus for long cable working, which, in his Siphon Recorder Patent, he says is as sensitive as his Mirror Galvanometer."

This principle, as the practical reader knows, has been largely used in telegraphy. Besides Lord Kelvin's application of it, we have the Brown and Allan Relay, the Weston Relay, and Voltmeter, and other contrivances of a similar nature;[1] but Mr Wilkins was himself the first to put it in practice, and under the following interesting circumstances: In 1851 he went to America to assist Henry O'Reilly of New York, a well-known journalist, who had a concession from the patentees of the Morse system for the erection of telegraph lines, at a royalty per mile. Disputes soon arose, and the Morse Syndicate sought to prevent O'Reilly from using their relay, without which the Morse instruments would be useless for long distances. In this difficulty O'Reilly adopted Bain's electro-chemical apparatus, and employed it for a time on the People's Telegraph from New York to Boston, *viâ* Albany. But finding that it was impossible to use this instrument in connection with intermediate stations, O'Reilly was again in a difficulty, when Mr Wilkins came to the rescue by saying he could devise a relay which did not require an iron armature, or electro-magnet of the ordinary form, and which would therefore be independent of the Morse patent. Very soon relays consisting of movable coils of wire, suspended between the poles of a magnet, were constructed in the workshop of John Gavitt, a friend of O'Reilly's, and then famous as a bank-note engraver. The instruments were placed in the

[1] The germ of all these instruments, as well as the Axial Magnets of Prof. Page and Royal E. House, was sown by Edward Davy in England in 1837. See my 'History of Electric Telegraphy,' pp. 356, 357.

circuit of the People's Telegraph, and O'Reilly was saved—but only for a time, as in the end he was beaten by his powerful opponents. The Wilkins relay was put aside and soon forgotten, but forty-three years later it was brought forward again by Mr Weston as an original invention.[1]

DR O'SHAUGHNESSY (AFTERWARDS SIR WILLIAM O'SHAUGHNESSY BROOKE)—1849.

One of the first difficulties encountered in the early days of the telegraph in India was the crossing of the great water-ways that abound in that country; and it was this difficulty which first directed the attention of Dr O'Shaughnessy, the introducer of the system in India, to the subject of subaqueous telegraphy.

In 1849 he laid a bare iron rod under the waters of the river Huldee, 4200 feet wide, with batteries and delicate needle instruments in connection on each bank. Signals were passed, but "it was found that the instruments required the attention of skilful operators, and that in practice such derangements occurred as caused very frequent interruptions."

He next tried the experiment without any metallic conductor, using the water alone as the sole vehicle of the electric impulses, but, though he again succeeded in passing intelligible signals, he found that the battery power for practical purposes would be enormous (he used up to 250 cells of the nitric acid and platinum form), and therefore prohibitively expensive.

Although for practical purposes he soon abandoned the idea of signalling across rivers with naked wires, and without any wires at all, O'Shaughnessy for many years took

[1] See the New York 'Electrical Engineer,' February 21, 1894.

great interest in the subject. Thus as late as 1858 we find him performing some careful experiments in the lake at Ootacamund, and in his Administration Report of the Telegraph Department for that year he says: "I have long since ascertained that two naked uncoated wires, kept a moderate distance—say 50 or 100 yards—apart, will transmit electric currents to considerable distances (two to three miles) sufficiently powerful for signalling with needle instruments."

E. AND H. HIGHTON—1852-72.

The brothers Edward and Henry Highton, who were well-known inventors in the early years of electric telegraphy, took up the problem of transaqueous communication about 1852. In Edward Highton's excellent little book, 'The Electric Telegraph: Its History and Progress,' published in that year, he says: "The author and his brother have tried many experiments on this subject. Naked wires have been sunk in canals, for the purpose of ascertaining the mathematical law which governs the loss of power when no insulation was used. Communications were made with ease over a distance of about a quarter of a mile. The result, however, has been to prove that telegraphic communications could not be sent to any considerable distance without the employment of an insulated medium."

On the other hand, Henry Highton long continued to believe in its practicability, and made many further experiments to that end. These were embodied in a paper read before the Society of Arts on May 1, 1872 (Telegraphy without Insulation), from which I condense the following account:—

"I have for many years been convinced of the possibility

of telegraphing for long distances without insulation, or with wires very imperfectly insulated; but till lately I had not the leisure or opportunity of trying sufficient experiments bearing on the subject. I need hardly say that the idea has been pronounced on all hands to be entirely visionary and impossible, and I have been warned of the folly of incurring any outlay in a matter where every attempt had hitherto failed. But I was so thoroughly convinced of the soundness of my views, and of the certainty of being able to go a considerable distance without any insulation, and any distance with very imperfect insulation, that I commenced, some three or four months since, a systematic series of experiments with a view to test my ideas practically.

"I began by trying various lengths of wire, dropped in the Thames from boats, and found that I could, without the slightest difficulty, exceed the limits allowed hitherto as practicable. This method, however, was attended with much difficulty and inconvenience, owing to the rapidity of the tides and the motion of the boats. I next tried wires across the Thames, but had them broken five or six times by the strength of the current and by barges dragging their anchors across them.

"I then put the instrument in my own room, on the banks of the river, and sent a boat down stream with a reel of wire and a battery to signal to me at different distances. The success was so much beyond my expectations, that I next obtained leave to lay down wires in Wimbledon Lake. As the result of all these experiments I found that water is so perfect an insulator for electricity of low tension that wires charged with it retained the charge with the utmost obstinacy; and, whether from the effect of polarisation (so-called), or, as I am inclined to suppose, from electrisation of the successive strata of water surrounding the wire, a long wire, brought to a state of low electrical tension, will retain that tension for minutes, or even hours. Notwithstanding attempts to discharge the wire every five seconds, I have found that a copper surface of 10 or 12 square feet in fresh water will retain a very appreciable charge for a quarter of an hour; and even when we attempt to discharge it continuously through a resistance of about thirty units [ohms], it will retain an appreciable though gradually decreasing charge for five or six minutes.[1]

"Since that time I have constructed an artificial line, consisting of resistance coils, condensers, and plates of copper in liquids, acting at once as faults and as condensers, so that I might learn as far as possible to what extent the principle of non-insulation can be carried, and I have satisfied myself that, though there are difficulties in very long lengths absolutely uninsulated, yet it is quite feasible to telegraph, even across the Atlantic, with an insulation of a single unit instead of the 170,000 units [absolute] of the present cables.

"The instrument with which I propose to work is the gold-leaf instrument, constructed by me for telegraphic purposes twenty-six years ago,[2] acted upon by a powerful electro-magnet, and with its motions optically enlarged. The exclusive use of this instrument in England was purchased by the Electric and International Telegraph Company, but it was never practically used, except in Baden, where a Government commission recommended it as the best. One of its chief merits is its extreme lightness and delicacy. Judging by the resistance it presents

[1] It does not appear to have struck our author that these effects would militate against the practical application of his method.

[2] A special arrangement of this instrument, adapting it for long and naked (or badly insulated) lines, was patented February 13, 1873. For reports of its great delicacy see 'Telegraphic Journal,' February 15, 1874.

to the electric current, it would appear that the piece of gold-leaf in the instrument now before us does not weigh more than $\frac{1}{2000}$th part of a grain; let us even say that it weighs four times more, or $\frac{1}{500}$th part of a grain. In order, then, to make a visible signal we only have to move a very, very small fraction of a grain through a very, very small fraction of an inch. You may judge of its delicacy when I show you that the warmth of the hand, or even a look, by means of the warmth of the face turned towards a thermopile, can transmit an appreciable signal through a resistance equal to that of the Atlantic cable (experiment performed). Another great merit of this instrument is its ready adaptability to the circumstances in which it may be placed, as it is easy to increase or diminish the length, or breadth, or tension of the gold-leaf. Thus, increase of length or diminution of breadth increases the resistance, but also increases the sensitiveness; and again, partaking as it does partly of the character of a pendulum and partly of a musical string, the rapidity of vibration is increased by giving it greater tension and greater shortness (though by doing so the sensitiveness is diminished), so that you can adjust it to the peculiar circumstances of any circuit. Again, you notice the deadness of the movements and the total absence of swing, which, whenever a needle is used, always more or less tends to confuse the signals. The greatest advantage of all is that we can increase the sensitiveness without increasing the resistance, simply by increasing the power of the electro-magnet.

"Having now explained the construction of the instrument, and pointed out its merits, I proceed to show by experiment how tenaciously a piece of copper in water will retain a state of electrical tension. Here is a tub of fresh water, with copper plates presenting to each other about

14 square feet of surface. I charge these plates with a Daniell cell, and you see how they retain the charge; in fact, they will go on gradually discharging for several minutes through the small resistance of the gold-leaf instrument. I now do the same with a tub of salt water, and the result is still the same, though less marked. In fact, these plates, with the water between, represent the two metallic surfaces of a Leyden jar, and the water retains the electricity of this small tension with much more obstinacy than the glass of a Leyden jar does the electricity of a higher tension.[1]

"Indeed, it is a fact of the highest importance in telegraphy that when there is a fault, electricity of a high tension, say of twenty or thirty Daniell cells, will almost wholly escape by it, and leave nothing for the instrument; whereas electricity of a small tension, as from a single cell of large surface, will pass through the instrument with very little loss of power. This is strikingly shown by the use of an ordinary tangent galvanometer. I cannot well show it to a large audience like the present, therefore I will only inform you that when I have taken two currents, each marking 30° on the galvanometer, the one of high tension from thirty Daniell cells, and the other of low tension from a single cell of small internal resistance, a fault equivalent to the exposure of a mile of No. 16 wire in sea-water will annihilate all appreciable effects on the galvanometer when using the current of high tension, whereas the current of low tension will still show as much as 20°. You see, then, the importance of using currents of low tension from a battery of large surface, and how a

[1] These experiments are not clearly described in the report from which we are quoting. If we understand them aright, they are rather electrolytic than Leyden-jar effects. In any case, as the tubs were presumably fairly well insulated, they have no bearing *ad rem.*

faulty cable can be worked with such currents when it is absolutely useless with currents of high tension.

"There are three ways of signalling without insulation: one, only feasible for short distances; a second, which I think will be found the most practicable; and a third, in the practical working of which for very long distances several difficulties (though by no means insuperable) present themselves.

"To explain the first plan, we will take the case of a river, and in the water near one bank place the copper

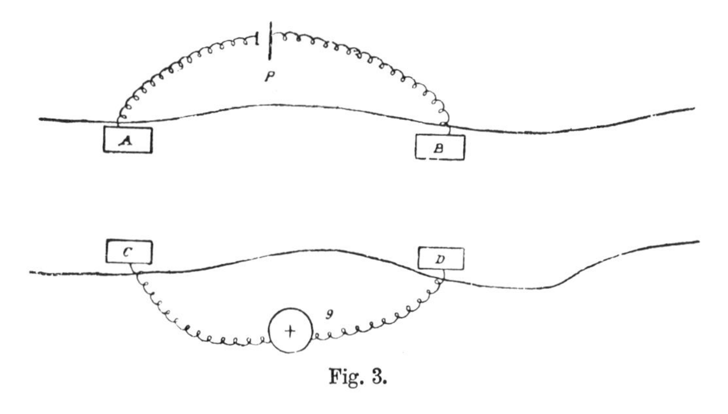

Fig. 3.

plates A B, and connect them with a wire, including the battery P. Near the opposite bank submerge similar plates, C D, connected by a wire, in the circuit of which is placed the galvanometer *g*. Between A and B the current will pass by every possible route, in quantities inversely proportional to their resistances; parts will pass direct by A B; and other portions by A, C, D, B, and by A, C, *g*, D, B. Now, if the plates be large, and A C and B D respectively comparatively near to each other, an appreciable current will pass from A to C, through *g*, and back from D to B; but if the plates be small, the battery power small, and the distance from A to B and from C to D comparatively short, no appreciable amount will pass through the galvanometer circuit. I do not hesitate to say that it is possible, by erecting a very thick line wire from the Hebrides to Cornwall, by the use of enormous plates at each extremity, and by an enormous amount of battery power—*i.e.*, as regards *quantity*—to transmit a current which would be sensibly perceived in a similar line of very thick wire, with very large plates, on the other side of the Atlantic. But the trouble and expense would probably be much greater than that of laying a wire across the ocean.

"The second is the simplest and most feasible plan—namely, laying across the sea two wires kept from metallic contact with each other, and working with that portion of the current which prefers to pass through this metallic circuit instead of passing across the liquid conductor, using currents of low tension from batteries of large surface.

"The third method is to lay a single wire imperfectly insulated, and to place at the opposite end beyond the instrument a very large earth-plate. Any electrical tension thrown on this wire transmits itself more or less to the opposite end, and will be shown on any instrument of small resistance and sufficient delicacy.[1] There are certain difficulties in this way of working, such as the effects of earth-currents and currents of polarisation which keep the needle or gold-leaf permanently deflected from zero, necessitating special means of counteraction. I have no doubt, from my experiments, that these difficulties may be overcome; but still I think the simplest and most feasible, and not more expensive, plan will be to work with two naked wires kept apart from metallic contact, using electricity of a very low tension."

[1] The following cutting from 'Once a-Week' (February 26, 1876) is given here in the hope that some American reader will kindly sup-

Soon after this Mr Highton turned a complete *volte face*, and went back to wires *perfectly* insulated, but at a *ridiculously* small cost! On April 20, 1873, he sent the following letter to the 'Times':—

"CHEAP TELEGRAPHY.

"SIR,—Some months ago I read a paper to the Society of Arts on the possibility of telegraphing for great distances without insulation, for which they were good enough to vote me a medal. I now find, however, that by the discovery of a new insulating material perfect insulation can be provided at a ridiculously small cost.

"I find by the addition of this material, which is simply tar chymically modified, nearly 200,000 per cent is added to the insulating power of a thin coating of gutta-percha. I hope the result will shortly be found in the great cheapening of telegraphy.—Yours, &c., H. HIGHTON."

The new material here referred to was a preparation of vegetable tar and oxide of lead, which almost instantly solidified on application. In some experiments at the Silvertown Works, it was found that No. 18 copper wire, covered with gutta-percha weighing only 21 lb. to the mile, had its insulation increased nearly 200,000 per cent, representing an insulation per mile of nearly three billion ohms!—enough, as the inventor needlessly remarked, for any lengths possible on the surface of the earth.[1]

ply details, if any are procurable: "The 'New York Tribune' gives an account of what appears to be a very remarkable discovery in electrical science and telegraphy. It is claimed that a new kind of electricity has been obtained, differing from the old in several particulars, and notably *in not requiring for transmission that the conducting wires shall be insulated.*"

[1] For reports on this cable see 'Telegraphic Journal,' vol. ii. pp.

G. E. DERING—1853.

The problem of wireless telegraphy was taken up about this time by Mr George Dering of Lockleys, Herts, who was, like his old Rugby tutor, Henry Highton, a prolific inventor of electrical and telegraphic appliances, patents for which he took out on eleven separate occasions between 1850 and 1858, and many of which came into practical use in the early Fifties. His needle telegraph, patented December 27, 1850, was in use in the Bank of England early in 1852, connecting the governor's room with the offices of the chief accountant, chief cashier, secretary, engineer, and other officials. About the same time it was partially used on the Great Northern Railway, and exclusively so on the first Dover-Calais cable (1851), where it did excellent service, working direct between London and Paris for a long time (including the busy period of the Crimean war), until supplanted by the Morse recording instrument.

In the same specification of 1850, Dering patented three methods of carrying off atmospheric electricity from the line-wires: (*a*) "Two roughened or grooved metallic surfaces separated by fine linen, one of which is included in the line-wire circuit, and the other is in connection with the earth." This was afterwards (in 1854) repatented by (Sir) William Siemens, and is now known as Siemens' Serrated-

104, 129. The Hightons received several Society of Arts' medals for the excellence of their telegraphic appliances which were largely used fifty years ago. Indeed a company, The British Electric Telegraph Co., was expressly formed in 1850 to work their instruments, and was afterwards merged in the British and Irish Magnetic Telegraph Co. A few years before his death (December 1874) Henry Highton invented an artificial stone, which I believe is largely used in building and paving.

Plate Lightning-Guard. (*b*) "The attraction or repulsion occurring between dissimilarly or similarly electrified bodies respectively. Thus metal balls may be suspended from the line-wire by wires, which on separating under the influence of the lightning-discharge make contact with plates connected with the earth; or the separation may simply break connection between the line-wire and the instrument." (*c*) "Introducing a strip of metallic leaf into the circuit, this being fused by the passage of the atmospheric electricity." This very effective method has also been reintroduced in later years, and always as a novelty, by various telegraph engineers.

Dering's telegraphic appliances made a goodly show at the Great Exhibition of 1851, side by side with Henley's colossal magnets, and received "honourable mention." They were again on view at the Paris International Exhibition of 1855, where they were awarded a medal for general excellence.

Dering's proposals for a transmarine telegraph are contained in his patent specification of August 15, 1853, from which we condense the following account:—

"The present invention is applicable to submarine telegraphs, and also to the means of communication by underground or over-ground wires. Heretofore, in constructing electric telegraphs where the whole circuit has been made of metal, and also where the conducting property of the earth has been employed as a part of the circuit, it has been usual, and it has been considered absolutely necessary, to cause the wires to be thoroughly insulated, the consequence of which has been that the expense of laying down electric circuits has been very great, particularly where the same have crossed the sea or other waters, where not only have the wires been insulated, but in order to protect the insulating matter from injury further great cost has been caused by the use of wire rope, or other means of protection.

"Now, I have discovered that a metallic circuit formed of wires, either wholly uninsulated or partially so, may be employed for an electric telegraph, provided that the two parts of the circuit are at such a distance apart that the electric current will not all pass direct from one wire to the other by the water or earth, but that a portion will follow the wire to the distant end.

"To carry out my invention, I cause two uninsulated or partially insulated wires to be placed in the water or in the earth, at a distance apart proportionate to the total length of the circuit, the said wires being insulated where they approach one another to communicate with the instruments, in order to prevent the current passing through the diminished water or earth space between them. The batteries (or other suitable source of electricity) employed are to be constructed in the proportion of their parts in conformity with the well-known laws which regulate the transmission of electric currents through multiple circuits—that is, they should possess the properties generally understood by the term *quantity* in a considerably greater degree than is usual for telegraphing through insulated wires, which may be effected (in the case of galvanic batteries) by using plates of larger dimensions, or by other alterations in the exciting liquids or plates. The proper distance at which to place the conductors from one another is also determined by the same laws, all of which will be readily understood by persons conversant with the principles of electrical science. In practice I find that from one-twentieth to one-tenth the length of the line-wires is a sufficient distance.

"Another method of carrying out my invention consists in establishing circuits composed in part of the uninsulated or partially insulated conductors, and in part of the conducting property of the sea, across which the communication is to be made, or of the earth or the moisture contained therein in the case of land telegraphs. For this purpose the connections are effected at such a distance in a lateral direction that a sufficient portion of the current will pass across the water or earth space and enter the corresponding wire connection at the other extremity. The connecting wires at the termini must be effectually insulated as in the first method.

"A third method consists in placing in the sea or earth two wires of dissimilar metal having the quality of generating electricity by the action of the water or moisture with which they are in contact. If at one extremity the wires be attached respectively to the two ends of the coil of an electro-magnet or other telegraphic apparatus, it will be found that the instrument is acted on by the current generated by the wires. If now at the other extremity the wires be connected, a portion of the current will complete its circuit through this connection, instead of all passing through the electro-magnet, where consequently the effect will be diminished; and if means be adopted to indicate this greater or less power, signals may be indicated at one end by making and breaking contact at the other. If desirable, currents derived from galvanic batteries, or other source, may be employed as auxiliary to those generated in the outstretched wires.

"In the different means of communication which I have described, if strong conductors are required, as in submarine lines, wire rope may be employed, either alone or attached to chains for greater strength and protection, or the conducting wires may be attached to hempen ropes, or enveloped within them. The metal composing the wires may be iron or copper or any other suitable kind, and it may be coated with varnish, by which means the amount of exposed surface will be diminished, and the metal preserved from corrosion.

"I will now suppose the case of a line to be carried out upon the principle which I have described, say from Holyhead to Dublin, a distance of about sixty miles. It would be necessary, first, to select two points on each coast from three to six miles apart, and to connect these points on each coast by insulated wires. Next, the two northern points are to be connected by a submerged uninsulated conductor, and the two southern points by a similar conductor, unless the water be employed as a substitute in the manner before described. Thus an oblong parallelogram of continuous conductors is formed, having for its longer sides the uninsulated conductors, and for its shorter sides the insulated wires along the coasts. If now these latter wires be cut at any parts, and instruments and batteries be connected in circuit, signals may be transmitted by any of the means ordinarily employed with insulated wires.

"Or, to take the case of a longer line, say from England to America, I should select two points, as the Land's End in Cornwall and the Giant's Causeway in Ireland or some suitable place on the west coast of Scotland, and corresponding points on the American shore. Next, I should unite the two points in each country by insulated wires, and, finally, submerge two uninsulated conductors across the Atlantic, or one if the water be employed to complete the circuit. Then by introducing, as before, suitable telegraphic instruments and batteries the communication will be established.

"From the foregoing description it will be seen that the cost of laying down electric telegraphs, whether submarine or otherwise, is, by this invention of employing distance between the conductors as a means of insulation, reduced to little more than the mere cost of the wires, together with that of an insulated wire at each end; while the numerous difficulties which attend the insulation of long lengths of wire are avoided, as also the chances of the communication being interrupted by accidents to the insulation."

At the time of this patent, and for many years after, the difficulties just referred to were only too real. Many of the cables laid between 1850 and 1860 failed after a longer or shorter period, and chiefly through defective insulation. Hence, no doubt, the persistency with which telegraph engineers in the Fifties sought in telegraphy without insulation, and telegraphy without wires, other and more economical ways of solving the great problem of trans-marine communication.

Dering's experiments were performed across the river Mimram at Lockleys, Herts, with bare parallel wires of No. 8 galvanised iron, laid at a distance apart of about 30 feet, or one-tenth of the space to be traversed. With a small battery power of only two or three Smee cells the signals were easily readable.

At one of these performances on August 12, 1853, the chairman and directors of the Electric Telegraph Company of Ireland (one of several mushroom companies then started) were present, and so impressed were they with the results obtained that they there and then decided to adopt the system for their intended line between Portpatrick and Donaghadee. This is a fact not generally known in the history of early submarine telegraph enterprises; and what is still less known, for there is no record of it, is that the project was actually attempted. In a recent letter, Mr Dering, who I am glad to say is still with us, has given me some interesting details of the attempt which I now publish, feeling sure that they will be new to the reader.

On September 23, 1853, the necessary wire in bundles was shipped to Belfast, which, "for the sake of ultra economy," consisted of single No. 1 galvanised iron instead of twisted strand wire as Dering had recommended. On examination the wire proved to be so unreliable, with numerous weak and brittle places—chiefly at the factory welds—that Dering urged delay and the substitution of stranded wire. "Had we been wise," writes Mr Dering, "we should have abandoned the attempt with this unsuitable material, but it was resolved to go on and risk it—testing the wire as far as might be beforehand and removing the weak parts. I, however, addressed a formal letter to the board of directors in London, stating that the wire was so unreliable I must decline all responsibility as to the laying it down, but that I would do the best I could."

After carefully testing the various lengths, removing all weak parts and bad welds as far as they could be discovered, and jointing and tarring the whole into one long length, the wire was paid into the hold of the Albert. On November 21 a start was made, a shore-end wire was laid from Milisle, carried out to sea, and buoyed. Next morning the Albert,[1] piloted by H.M.S. Asp (Lieut. Aldridge), picked up the buoyed end, joined it to the wire on board, and paid out successfully for about 3½ miles, when the wire broke at a factory weld, and the ship returned to Donaghadee "in a gale of wind."

The next few days were occupied in some alterations to the paying-out machinery, found by experience to be desirable, and on the 26th another start was made. The wire on board was joined to the buoyed end at 4 miles from shore, and paying-out proceeded successfully as far as mid-channel (about 12 miles) when the wire broke, again

[1] With Dr Hamel on board, the famous Russian scientist of Alpine celebrity, as the representative of his Government.

at a factory weld, and the end was lost in 82 fathoms of water. The ship then returned to the buoy and tried to underrun the wire, but it soon broke again, and for the moment further attempts were abandoned.

Previous to this two unsuccessful attempts had already been made to connect Great Britain and Ireland by cables made on the lines of the Dover-Calais cable of 1851—one, undertaken by Messrs Newall & Co., between Holyhead and Howth, June 1, 1852, which failed three days after; and the other, a heavy six-wired cable, undertaken by the same firm, between Portpatrick and Donaghadee, October 9, 1852, which broke in a gale after sixteen miles had been paid out.

In June 1854 Messrs Newall recovered the whole of this sixteen miles of cable, and completed the laying to Portpatrick, thus rendering another attempt at a bare wire cable unnecessary, if, indeed, it was still thought desirable.

Mr Dering's faith in the soundness of his views is still unshaken, for he goes on to say: "Instead of a single wire, as in 1853, I should now advocate the use of a bare strand of wires for each of the conductors. And I must add, considering the craving there is at present for Wireless Telegraphs, that it seems to me not altogether improbable that the less ambitious but (for, at all events, long distances) far more feasible plan of using bare wires will yet have its innings." And who, in these days of electrical marvels, will dare to say him nay? I, for my part, will not, for I have seen more unlikely things come to pass. The dream of to-day, "idle and ridiculous" as it may seem, has been so often realised on the morrow, that the cautious historian of science must not look for finality in any of its applications.[1]

[1] For recent applications of the bare-wire principle, see Melhuish, p. 111, *infra*.

JOHN HAWORTH—1862.

On March 27, 1862, Mr Haworth patented "An improved method of conveying electric signals without the intervention of any continuous artificial conductor," in reference to which a lecturer of the period said:[1] "I have not met one single gentleman connected with the science of telegraphy who could understand his process, or its probability of success. I applied to him for some information, but he is unwilling to communicate any particulars until experiment has sufficiently demonstrated the practicability of his plans."

In the discussion which followed, Mr Cromwell Varley, electrician of the old Electric and International Telegraph, and the old Atlantic Telegraph, Companies, said: "Being informed that Sir Fitzroy Kelly and the learned chairman (Mr Grove) had seen Haworth's system in operation, and that the latter gentleman was a believer in it, he had tried the experiment upon a very small scale in his own garden, with apparatus constructed according to the instructions of Mr Haworth. His two stations were only 8 yards apart, and, although he used a very sensitive reflecting galvanometer, and twelve cells of Grove's nitric acid battery, he could not get any signals, although the experiments were varied in every conceivable way."

Under these circumstances it will not be surprising if I, too, after a careful study of the specification, and with the light thrown upon it by a further patent of October 30, 1863, have failed to understand the author's method. Indeed, I feel in much the same mental condition towards it as Tristram Shandy's connoisseurs, who, "by long friction, incumbition, and electrical assimilation, have the happiness,

[1] T. A. Masey, Society of Arts, January 28, 1863.

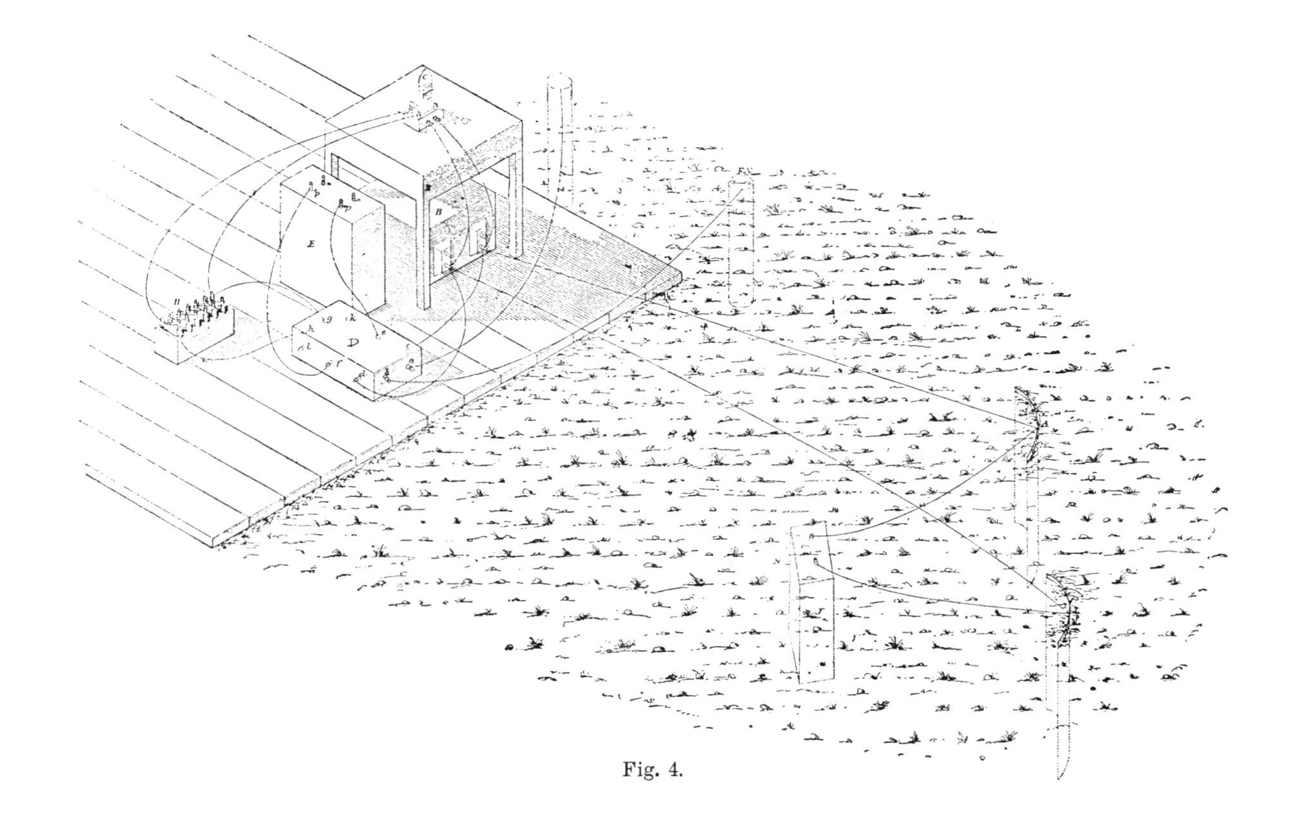

Fig. 4.

at length, to get all be-virtu'd, be-pictured, be-butterflied, and be-fuddled." However, I will do my best to translate the terrible phraseology of the letters patent into plain English; and if after this my readers cannot divine the mode of action I will not blame them—nor must they blame me! My description of the apparatus is based on the complete specification and drawings of the second patent, which were lodged in the Patent Office on April 30, 1864, and which must therefore be supposed to contain the inventor's last word on the subject.

A, Z (fig. 4) are copper and zinc plates respectively, curved as shown, and buried in the earth about 3 feet

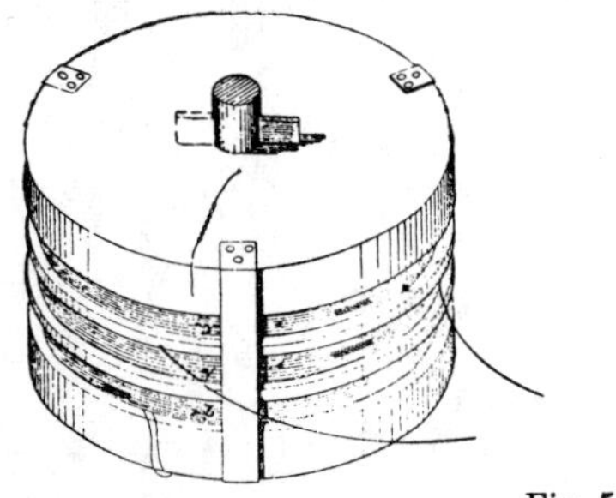

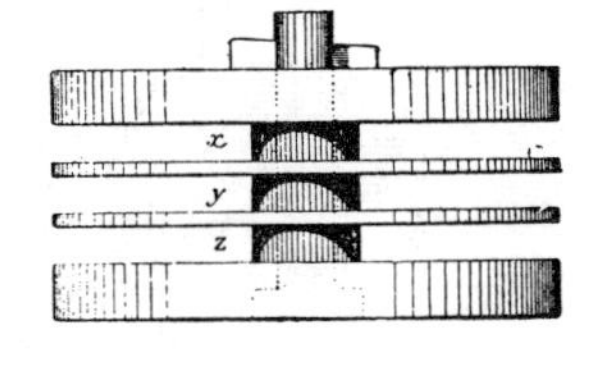

Fig. 5.

apart. The superficies varies according to distance and other circumstances: thus, for distances up to 75 miles plates 1 foot square suffice; over 75 and up to 440 miles, plates 24 by 16 inches are required. G, F are copper cylinders, 24 by 4 inches, buried in earth, which is always moist. At a point distant about 3 feet from the centres of A and Z a wooden box J is buried, containing a coil of insulated copper wire, No. 16 gauge, wound upon a wooden reel. The ends of the coil are attached to binding screws shown on top of the box. B is a wooden box containing a wooden reel divided into three compartments, *x*, *y*, *z* (fig. 5). *x* is filled with fine covered-copper wire, the ends of which are brought together and secured on the outside of the reel. *y* is filled with thicker covered-copper wire, wound in the same direction as *x*, and the ends are severally connected to binding-screws, shown on the outside. *z* is half filled with insulated iron wire, wound in the same direction as *x* and *y*; the ends are fastened together on the outside of the reel as with coil *x*. The compartment is then filled with more of the same iron wire, wound double, and in the reverse direction to the coil below it. These double wires are not twisted, nor bound together, nor allowed to cross one another, but are wound evenly in layers side by side; and the ends of each coil are secured together on the outside of the reel as in the case of the lower coil, and adjacent thereto. Usually the wire of coil *x* is No. 32 gauge; *y*, No. 16; and *z*, No. 20; but the sizes and quantities required must vary according to distance and other circumstances.

C is any suitable telegraph instrument of the needle pattern.

D is a condenser of a kind which an electrical Dominie Sampson would call prodigious! A wooden box divided lengthwise into two compartments well coated with shellac. In each compartment is placed a band of stout gold-foil—both well insulated, and connected at their ends to the binding-screws *a*, *g*, and *b*, *h*, respectively (fig. 6). Each compartment is filled with sixty rectangular plates of gutta-percha, on which insulated copper wire, No. 32 gauge, is wound in one continuous length from the first plate to the last, and the ends are attached to the binding-screws *a*, *g*, and *b*, *h*, respectively. "I fix binding-screws *c*, *d*, *e*, *f*, *k*, and *l* in the positions shown, and connect them with the wire upon the plates in its passage through the box. I then pass from end to end of each compartment over the plates, and lying on them, but well insulated from them, another band of stout gold-foil, and connect each end of it with the screws *a*, *g*, and *b*, *h*, respectively."

E is another wooden box, containing a reel similar to B, but divided into only two compartments, each of which is filled with two copper wires, one covered and the other uncovered, wound side by side, and all four of different gauges from No. 18 to 30. The ends of one of the covered coils are brought to the screws *p*, *p*, shown on top of the box; the ends of the other covered coil are fastened

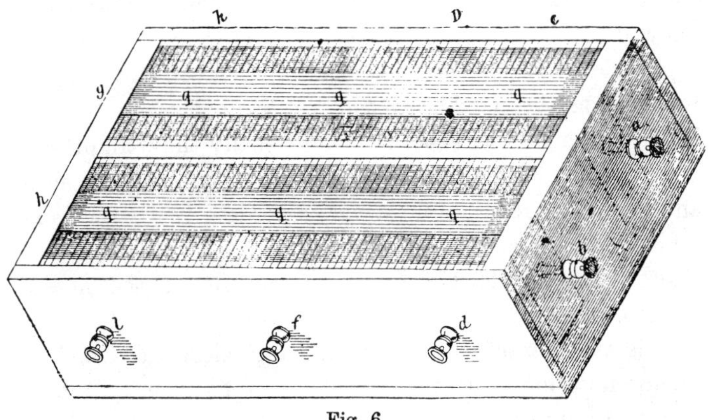

Fig. 6.

on the outside of the reel; and the ends of the two uncovered coils are likewise fastened on the outside of the reel, "but in such a position that they can never come in contact with any uncovered part of the coated wire. Between each of the layers of wire I place a strip of non-metallic paper to insulate it from the layers above and below, and when in winding I arrive within an inch of the circumference of the reel I employ gutta-percha tissue in addition to the non-metallic paper."

H is a Smee's battery, the size and power of which will depend on circumstances, such as the distance to which it is intended to convey the message; the strength and direction of earth-currents; and even the state of the weather—more power being required in dry than in damp weather. "For a distance of ten miles, from Notting Hill to Croydon, I have found a Smee's battery of two cells at each end, containing plates 3 by 5 inches, to suffice. For about fifty miles, from Notting Hill to Brighton, I have used with success a battery of three cells at each end; and from Notting Hill to Bangor, in Wales, I have required six cells at each end. Generally speaking, I have found that less power is required to convey a message from north to south and from south to north than from east to west, or from west to east."

The connections of the various instruments are shown by lines, and an exactly similar set of instruments is arranged at the place with which it is desired to correspond.

And now as to the *modus operandi:* when the handle of the needle instrument, C, is worked in the act of signalling, what happens? Here the trouble comes in. The author, I regret to say, is silent as to what happens, and I won't be so rash as to make a guess; but I would suggest the question as a safe prize-puzzle for the Questions and Answers column of some technical journal! Seriously, it seems to me that the results, if any, must be a perfect chaos of battery currents, earth-battery currents, earth-currents, induction currents, and currents of polarisation —all fighting in a feeble way for the mastery; and yet some men, besides the author, believed these effects to be intelligible signals!

The remarks of Mr Varley, quoted above, drew that gentleman into an angry correspondence in the pages of the old 'Electrician' journal, from which I need only extract Mr Varley's letter and Mr Haworth's reply. In the number for February 27, 1863, Mr Varley writes:—

"I make it a rule never to pay any attention to anonymous correspondents. As Mr Haworth, however, has commented upon the remarks I made a short time since at the Society of Arts, allow me to draw attention to the fact that, the discussion having been prolonged beyond the time allotted for that purpose, the detail of the experiments could not then be fully entered into.

"Mr Haworth paid me 'one' visit a short time ago, when I asked him if he had any objection to his invention being tested by actual experiment: he said he had not, and pointed out to me how to arrange the various parts of the apparatus. I have preserved the pencil sketch made at the time, as indicated and approved by him. This was strictly followed in the experiments.

"The apparatus used was constructed especially for this purpose. The primary coils were *thoroughly* insulated with gutta-percha, the secondary coils by means of a resinous compound and india-rubber. The plates of copper and zinc at each station were but an inch and a half from each other; they were each 6 inches square. The two stations were only 8 yards apart.

"The apparatus at each station consisted of a plate of copper and a plate of zinc, connected to a flat secondary coil containing nearly a mile of No. 35 copper wire. The secondary coil was placed immediately behind the plates, and behind this was placed a flat primary coil.

"At the sending station the primary coil was connected with six cells of Grove's battery, and contact intermitted. At the receiving station the primary coil was connected with one of Thomson's reflecting galvanometers, of small resistance, no more than that of an ordinary telegraph instrument.

"With this disposition of apparatus *no current* could be obtained.

"Crossing a river without wires is an old experiment. In March 1847 I tried experiments in my own garden, and also across the Regent's Canal, with a single cell of Grove's battery. Feeble but evident currents were sent across the canal 50 feet wide. The current received was but a minute fraction of that leaving the battery. In this case the distance across the canal was but one quarter of that separating the plates on each bank. When, however, these plates were brought near together, as in Haworth's specification, no visible signal could be obtained.

"This experiment has been repeated by numbers in various parts of the world, and with the same well-known results. When tried by me in 1847, I was unaware that the idea had occurred to Professor Morse, or any one else.

"To account for Mr Haworth's assertions that he has worked from Ireland to London, and between other distant places, I can only suppose that he has mistaken some irregularity in the currents generated by his copper and zinc plates for signals.[1]

"If he *can* telegraph without wires, why does he not connect England with America, when he can earn £1000 per diem forthwith, and confer upon the world a great blessing?

"Before speaking at the Society of Arts, I called at Mr Haworth's house several times, and found him out on all occasions. I wrote him more than once, giving him the negative results of my experiments, &c. He, however, paid no attention to any of my communications.

[1] "I have seen Mr Haworth's apparatus at work repeatedly, and have myself read off from the indicator the messages which have arrived; and these 'irregular currents mistaken for signals' have consisted of words and sentences transmitted as correctly as by the electric telegraph. My house has been one station, and Brighton, or Kingstown in Ireland, the other."—J. M. Holt, 'Electrician,' March 6, 1863.

"I have not been able to meet with a single individual who has seen a message transmitted by Mr Haworth; and every one of those who are reported to have seen it, and with whom I have come in contact, positively deny it when questioned.

"I have no hesitation in stating—1st, That Mr Haworth's specification is unintelligible: it is a jumble of induction plates, induction coils, and coils of wire connected together in a way that can have no meaning.

"2ndly, That he cannot send electric signals without wires to any useful distance.

"3rdly, From my acquaintance with the laws of electricity, I cannot believe it possible that he has ever communicated between distant stations as stated in his specification, No. 843, 1862.

"4thly, Supposing for a moment that he could work, as stated, any person constructing a similar apparatus in the neighbourhood would be able to read the communications, and they no longer would be private."

In the following number (March 6, 1863) Mr Haworth says:—

"Will you kindly allow me space for a line in reply to Mr Varley? I never received his letter of the 27th of January, and am truly sorry for any apparent discourtesy on my part. I fear other letters have shared the same fate.

"From Mr Varley's account of his experiments I find several particulars in which there has been considerable misapprehension on his part; but I cannot spare the time —nor can I ask you for the space—to give further explanations. It certainly is a new feature in electricity, if the earth's currents alone can register words and sentences on the dial-plate. I hope shortly to be able to convince the most sceptical by ocular demonstration. For the present I am content to wait, being anxious rather to perfect my discovery than to push it."

After this we hear nothing more of Mr Haworth, though no doubt the publication and discussion of his views kept the subject alive for a time.[1]

J. H. MOWER—1868.

Of the next proposal with which we have to deal in these pages, I find amongst my notes only a single cutting from the New York 'Round Table' of (August or September) 1868. I give it, *in extenso*, for what it is worth, and hope some American reader may be able to furnish details and further developments if any:—

"Mr Mower has elaborated a discovery which, if the description given by the 'New York Herald' is to be relied upon, will revolutionise trans-oceanic, and generally all subaqueous, telegraphy. For some years he had been engrossed in electrical experiments, when the Atlantic cable gave a special direction to his investigations into generating and conducting substances, the decomposition of water, the development of the electrical machine, &c., &c. By this summer his arrangements had been so far perfected that, a few weeks ago, he was able to demonstrate to himself and his coadjutor the feasibility of his project, on a scale approximate to that which it is designed to assume.

"Selecting the greatest clear distance on an east and west line in Lake Ontario—from a point near Toronto,

[1] See, for example, 'The Electrician,' January 23, 1863. Also Béron's 'Météorologie Simplifiée,' Paris, 1863, pp. 936, 937, where there is a hazy description of a wireless telegraph, apparently on the same lines as Haworth's.

Canada West, to one on the coast of Oswego County, New York—at his first attempt he succeeded in transmitting his message, without a wire, from the submerged machine at one end of the route to that at the other. The messages and replies were continued for two hours, the average time of transmission for the 138 miles being a little less than three-eighths of a second.

"The upshot of the discovery—on what principle Mr Mower is not yet prepared to disclose—is, that electric currents can be transmitted through water, salt or fresh, without deviation vertically, or from the parallel of latitude. The difficulty from the unequal level of the tidal waves in the two hemispheres will be obviated, it is claimed, by submerging the apparatus at sufficient depth. The inventor, we are told, is preparing to go to Europe to secure there the patent rights for which the caveats have been filed here. At the inconsiderable cost of 10,000 dollars he expects within three months to establish telegraphic communication between Montauk Point, the eastern extremity of Long Island, and Spain, the eastern end of the line striking the coast of Portugal at a point near Oporto.

"The statement of the discovery is enough to take away one's breath; but, with the history of the telegraph before us, we no more venture to deny than we do to affirm its possibility."

M. BOURBOUZE—1870.

During the investment and siege of Paris by the German forces in the winter of 1870-71, many suggestions were made for the re-establishment of telegraphic communication between Paris and the provinces. Acoustic methods were tried, based on the transmission of sound by earth and water. A Mr Granier proposed a form of aerial line which was thought to be feasible by the distinguished aeronaut, Gaston Tissandier. The wire (to be paid out from balloons) was to be enclosed in gutta-percha tubing, inflated with hydrogen gas so as to float 1000 to 1500 metres above the earth.[1]

Amongst other suggestions was one by M. Bourbouze, a well-known French electrician, which only need concern us in these pages. His proposal was to send strong currents into the river Seine from a battery at the nearest approachable point outside the German lines, and to receive in Paris through a delicate galvanometer such part of these currents as might be picked up by a metal plate sunk in the river. After some preliminary experiments between the Hotel de Ville and the manufactory of M. Claparède at St Denis, it was decided to put the plan in practice. Accordingly, on December 17, 1870, M. d'Almeida left the beleaguered city by balloon, descended after many perils at Champagne outside the enemy's lines, and proceeded *viâ* Lyons and Bordeaux to Havre. Thence the necessary apparatus was ordered from England and conveyed to Poissy, where M. d'Almeida regained the banks of the Seine on January 14, 1871. Here, however, the river was found to be completely frozen over, and the attempt at communicating with Paris was deferred to January 24. Meanwhile the armistice was proclaimed, and the project was allowed to drop.[2]

[1] Such a plan was patented in England more than twenty years previously. See patent specification, No. 2907, of November 19, 1857.

[2] On March 27, 1876, Bourbouze requested to be opened at the Academy of Sciences a sealed packet which he had deposited on November 28, 1870. It was found to contain a note entitled "Sur les Communications a Distance par les Cours d'Eau." The contents of the document, so far as I know, have not been published.

M. Bourbouze did not, however, abandon his idea, and, thinking he found in the principle of La Cour's phonic wheel telegraph a better means of indicating the signals than the galvanometer, he again took up the problem. Between 1876 and 1878 an occasional notice of his experiments appeared in the technical journals, but they are all provokingly silent on the point of actual results over considerable distances.[1]

MAHLON LOOMIS—1872.

In 1872 Mr Mahlon Loomis, an American dentist, proposed to utilise the electricity of the higher atmosphere for telegraphic purposes in a way which caused some excitement in America at the time.

It had long been known that the atmosphere is always charged with electricity, and that this charge increases with the ascent: thus, if at the surface of the earth we represent the electrical state or charge as 1, at an elevation of 100 feet it may be represented as 2; at 200 feet as 3; and so on in an ascending series of imaginary strata. Hitherto this had been considered as a rough-and-ready way of stating an electrical fact, just as we say that the atmosphere itself may, for the sake of illustration, be divided into strata of 100 or any agreed number of feet, and that its density decreases *pro rata* as we ascend through each stratum. But Mr Loomis appears to have made the further discovery that these electrical charges are in some way independent of each other, and that the electricity of any one stratum can be drawn off without the balance being

[1] See, amongst other accounts, the 'English Mechanic,' September 8, 1876; 'Engineering,' April 13, 1878; and the French journal, 'La Nature,' July 8, 1876. For Bourbouze's earlier experiments, see 'La Lumière Électrique,' August 19, 1879.

immediately restored by a general redistribution of electricity from the adjacent strata. On this assumption, which is a very large one, he thought it would be easy to tap the electricity at any one point of a stratum, preferably an elevated one where the atmosphere is comparatively undisturbed, which tapping would be made manifest at any distant point of the same stratum by a corresponding fall or disturbance there of the electrical density; and thus, he argued, an aerial telegraph could be constructed.

The following is an extract from his (American) patent, dated July 30, 1872:—

"The nature of my discovery consists in utilising natural electricity, and establishing an electrical current or circuit for telegraphic and other purposes without the aid of wires, artificial batteries, or cables, and yet capable of communicating from one continent of the globe to another.

"As it was found possible to dispense with the double wire (which was first used in telegraphing), making use of but one, and substituting the earth instead of a wire to form the return half of the circuit; so I now dispense with both wires, using the earth as one-half the circuit and the continuous electrical element far above the earth's surface for the other half. I also dispense with all artificial batteries, but use the free electricity of the atmosphere, co-operating with that of the earth, to supply the current for telegraphing and for other useful purposes, such as light, heat, and motive power.

"As atmospheric electricity is found more and more abundant when moisture, clouds, heated currents of air, and other dissipating influences are left below and a greater altitude attained, my plan is to seek as high an elevation as practicable on the tops of high mountains, and thus establish electrical connection with the atmospheric stratum or ocean overlying local disturbances. Upon these mountain-tops I erect suitable towers and apparatus to attract the electricity, or, in other words, to disturb the electrical equilibrium, and thus obtain a current of electricity, or shocks or pulsations, which traverse or disturb the positive electrical body of the atmosphere between two given points by connecting it to the negative electrical body of the earth below."

To test this idea, he selected two lofty peaks on the mountains of West Virginia, of the same altitude, and about ten miles apart. From these he sent up two kites, held by strings in which fine copper wires were enclosed. To the ground end of the wire on one peak he connected an electrical detector—presumably of the electrometer kind—and on the other peak a key for connecting the kite wire to earth when required. With this arrangement we are told that messages were sent and received by making and breaking the earth connection, "the only electro-motor being the atmospheric current between the kites, and which was always available except when the weather was violently broken."

So well did this idea "take on" in the States that we learn from the New York 'Journal of Commerce' (February 5, 1873) that a bill had passed Congress incorporating a company to carry it out. The article then goes on to say: "We will not record ourselves as disbelievers in the Aerial Telegraph, but wait meekly and see what the Doctor will do with his brilliant idea now that both Houses of Congress have passed a bill incorporating a company for him. Congressmen, at least, do not think him wholly visionary; and it is said that the President will sign the bill; all of which is some evidence that air telegraphy has another side than the ridiculous one. The company receive no money from the Government, and ask none. As we understand the Loomis plan, it is something to this effect—and readers are cautioned not to laugh too boisterously at it, as also not to believe in it till demonstrated. The inventor proposes to build a very tall tower on the highest peak of the Rocky Mountains. A mast, also very tall, will stand on this tower, and an apparatus for 'collecting electricity' will top the whole. From the loftiest peak of the Alps will rise another very tall tower and ditto mast, with its coronal electrical affair. At these sky-piercing heights Dr Loomis contends that he will reach a stratum of air loaded with electricity; and we cannot say that he will not. Then, establishing his ground-wire connections the same as in ordinary telegraphs, he feels confident that he can send messages between the mast-tops, the electrified stratum of air making the circuit complete. The inventor claims to have proved the feasibility of this grand scheme on a small scale. We are told that, from two of the spurs of the Blue Ridge Mountains, twenty miles apart, he sent up kites, using small copper wire instead of pack-thread, and telegraphed from one point to the other."

At intervals in the next few years brief notices of the Loomis method appeared in the American journals, some of which were copied into English papers. The last that I have seen is contained in the 'Electrical Review' of March 1, 1879, where it is stated that "with telephones in this aerial circuit he [Loomis] can converse a distance of twenty miles," to which the editor significantly adds a note of interrogation.

The fact is, however much Mr Loomis and his Wall Street friends believed that dollars were in the idea, the technical press never took it very seriously. This is shown by the following cutting, which we take from the New York 'Journal of the Telegraph,' March 15, 1877: "The never-ending procession of would-be inventors who from day to day haunt the corridors and offices of the Electrician's department at 195 Broadway, bringing with them mysterious packages tied up in newspapers, was varied the other day by the appearance of a veritable lunatic. He announced that that much-talked-of great discovery of a few years ago, aerial telegraphy, was in actual operation right here in New York. A. M. Palmer, of the Union Square Theatre, together with one of his confederates, alone possessed the secret! They had unfortunately chosen to use it for illegitimate purposes, and our visitor, therefore, felt it to be his solemn duty to expose them. By means of a $60,000 battery, he said, they transmitted the subtle fluid through the aerial spaces, read people's secret thoughts, knocked them senseless in the street; ay, they could even burn a man to a crisp, miles and miles away, and he no more know what had hurt him than if he had been struck by a flash of lightning, as indeed he had![1] The object of our mad friend in dropping in was merely to ascertain how he could protect himself from Palmer's illegitimate thunderbolts. Here the legal gentleman, lifting his eyes from 'Curtis on Patents,' remarked: 'Now, I'll tell you what you do. Bring a suit against Palmer for infringement of Mahlon Loomis's patent. Here it is' (taking down a bound volume of the 'Official Gazette'), 'No. 129,971. That'll fix Palmer.'"

In conclusion of this period of our history, it will suffice to say that between 1858 and 1874 many patents were taken out in England for electric signalling on the bare wire system of Highton and Dering, with or without the use of the so-called "earth battery." As they are all very much alike, and all unsupported, so far as I have seen, by any experimental proofs, it would be a tiresome reiteration to describe them, even in the briefest way. I therefore content myself with giving the following list, which will

[1] This lunatic must be still abroad, for we occasionally hear much the same thing of the diabolic practices of Tesla and Marconi.

be useful to those of my readers who desire to consult them.

Name of patentee.	No. and date of patent.	
B. Nickels . . .	2317	October 16, 1858.
A. V. Newton . .	2514	November 9, 1858.
A. Barclay . . .	56	January 7, 1859.
Do. . . .	263	January 28, 1859.
J. Molesworth . .	687	March 18, 1859.
H. S. Rosser . .	2433	October 25, 1859.
W. E. Newton . .	1169	May 11, 1860.
H. Wilde . . .	2997	November 28, 1861.
Lord A. S. Churchill .	458	February 20, 1862.
H. Wilde . . .	3006	December 1, 1863.
Do. . . .	2762	October 26, 1865.
T. Walker . . .	2870	November 6, 1866.
Do. . . .	293	January 23, 1874

SECOND PERIOD—THE PRACTICABLE.

PRELIMINARY: NOTICE OF THE TELEPHONE IN RELATION TO WIRELESS TELEGRAPHY.

"Give me the ocular proof,
Make me see't; or, at least, so prove it,
That the probation bear no hinge, nor loop,
To hang a doubt on."

We have now arrived at a period in the history of our subject at which experiments begin to assume a character more hopeful of practical results. All that went before was more or less crude and empirical, and could not be otherwise from the very necessities of the case. The introduction of the telephone in 1876 placed in the hands of the electrician an instrument of marvellous delicacy, compared with which the most sensitive apparatus hitherto employed was as the eye to the eye aided by the microscope. Thus, Prof. Pierce of Providence, Rhode Island, has found that the Bell telephone gives audible signals with considerably less than the one-hundred-thousandth part of the current of a single Leclanché cell. In testing resistances with a Wheatstone bridge, the telephone is far more sensitive than the mirror galvanometer; in ascertaining the continuity of fine wire coils it gives the readiest answers; and for all the different forms of atmospheric electrical discharges—and they are many—it has a language of its own, and opens up to research a new field in meteorology.

The sound produced in the telephone by lightning, even when so distant that only the flash can be seen in the horizon, and no thunder can be heard, is very characteristic—something like the quenching of a drop of molten metal in water, or the sound of a distant rocket; but the remarkable circumstance for us in this history is, that this sound is always heard just *before* the flash is seen, showing that there is an inductive disturbance of the electricity *overhead*, due to the *distant* concentration preceding the disruptive discharge. Thus, on November 18, 1877, these peculiar sounds were heard in Providence, and the papers next morning explained them by reporting thunderstorms in Massachusetts. Sounds like those produced by lightning, but fainter, are almost always heard many hours before a thunderstorm actually breaks.[1]

The Bell telephone was tried for the first time on a wire from New York to Boston on April 2, 1877, and soon afterwards its extraordinary sensitiveness to induction currents, and currents through the earth (leakages) from distant telegraph circuits, began to be observed.[2] Thus, in August 1877, Mr Charles Rathbone of Albany, N.Y., had been experimenting with a Bell telephone which was attached to a private telegraph line connecting his house with the Ob-

[1] 'Journal of the Telegraph,' N.Y., December 1, 1877. See also 'Jour. Inst. Elec. Engs.,' vol. vi. p. 523, vol. vii. p. 329; 'The Electrician,' vol. ix. p. 362.

[2] The disturbing effects of induction on ordinary telegraph wires on the *same* poles had long before this been noticed. See Culley's paper and the discussion thereon in the 'Jour. Inst. Elec. Engs.,' vol. iv. p. 54. See also p. 427 for Winter's interesting observations in India in 1873. As far back as 1868 Prof. Hughes, at the request of the French Telegraph Administration, undertook a series of experiments with a view of finding a remedy. The results are given in his paper read before the Inst. Elec. Engs., March 12, 1879.

servatory. One evening he heard some singing which he thought came from the Observatory, but found on inquiry that that was not the case. He then carefully noted what followed, and next morning sent a note to the newspapers stating the facts and giving the names of the tunes which he had heard. This elicited the information that the tunes were those of an experimental concert with Edison's singing telephone over a telegraph wire between New York and Saratoga Springs. It was then resolved to follow up this curious discovery, and, accordingly, when Edison's agent gave another concert in Troy, arrangements were made to observe the effects. A wire running from Albany to Troy alongside the Edison wire was earthed with a Bell telephone in circuit at each end. The concert was heard as before, the music coming perfectly clear, and the tunes distinguishable without the least difficulty.

Later in the evening the instruments were put in circuit on one of the wires running from Albany to New York. Again the music was heard, and much louder, so that by placing the telephone in the centre of the room persons seated around could hear with perfect distinctness.

These observations were made on six separate occasions between August 28 and September 11, and, strangely enough, two other and independent observers in Providence, 200 miles away, noted the same effects on five out of the six dates given by Mr Rathbone.[1]

Dr Channing, one of the observers in Providence, has published a very interesting account[2] of his observations, from which I will make a few extracts. During five

[1] 'Journal of the Telegraph,' N.Y., October 1 and 16, and November 1, 1877. For other early observations of the same kind see 'The Telegraphic Journal,' March 1, 1878, p. 96 ; 'Journal of the Telegraph,' March 16, 1878 ; 'The Electrician,' vol. vi. pp. 207, 303.

[2] 'Journal of the Telegraph,' December 1, 1877, and reproduced in the 'Jour. Inst. Elec. Engs.,' vol. vi. p. 545.

evenings in the latter part of August and first part of September 1877 concerts were given in the Western Union Office, N.Y., for the benefit of audiences in Saratoga, Troy, and Albany respectively. The performers sang or played into an Edison musical telephone, actuated by a powerful battery, and connected with one or other of the above-named places by an ordinary telegraph line, with return through the ground.

In Providence, on the evening of the first concert, Dr Channing and a friend were conversing through Bell telephones over a shunt wire, made by grounding one of the American District Telegraph wires at two places, a quarter of a mile apart, through the telephones and several hundred ohms resistance. At about half-past eight o'clock they were surprised by hearing singing on the line, at first faint, but afterwards becoming clear and distinct. Afterwards, during that and subsequent evenings, various airs were heard, sung by a tenor or soprano voice, or played on the cornet. On investigation, the music heard proved to be the same as that of the Edison concerts performed in New York.

The question how this music passed from the New York and Albany wire to a shunt on the District wire in Providence is of scientific importance. The Edison musical telephone consists of an instrument which converts sound waves into galvanic waves at the transmitting station, and another apparatus which reconverts galvanic waves into sound waves at the receiving station. The battery used in these concerts consisted of 125 carbon-bichromate cells (No. $1\frac{1}{2}$), with from 1000 to 3000 ohms resistance interposed between the battery and the line. The line wire extended from the Western Union office, *viâ* the Harlem Railway, to Albany. On the same poles with this Albany wire, for sixteen miles, are carried four other wires, all running to Providence, and also, for eight miles, a fifth wire from Boston, *viâ* New London, to Providence. All these lines, including the Albany wire, are understood to have a common earth connection at New York, and to be strung at the usual distance apart, and with the ordinary insulation.

At Providence six New York and Boston wires run into the Western Union office on the same poles and brackets for the last 975 feet with an American District wire. This wire belongs to an exclusively metallic circuit of four and a half miles, having, therefore, no earth connection. Finally, in a shunt on this wire, the telephones were placed as before described.

It will thus be seen that the music from the Albany wire passed first to the parallel New York–Providence wires; secondly, from these to a parallel District wire in Providence; and thirdly, through a shunt on the District wire to the telephones.

This transfer may have taken place by induction, by cross-leakage, or, in the first instance, in New York by a crowded ground connection; but in the transfer in Providence from the New York–Boston to the District wire there was no common ground connection, and it is difficult to suppose that sufficient leakage took place on the three brackets and three poles (common to the New York and District wires) to account for it. Without wholly rejecting the other modes of transfer, Dr Channing ascribes to induction the principal part in the effects.

The next question arises, What proportion of the electrical force set in motion in New York could have reached the listeners on the short shunt line in Providence? Whether induction or cross-leakage or crowded ground was concerned, who will say that the New York–Providence wires had robbed the Albany wire of one-tenth or even one-hundredth of its electrical force? When this reached Providence, did the New York wires in the course of 975 feet give up to the District wire one-tenth or one-hundredth of *their* force? Lastly, when the District circuit had secured this minute fraction, did the shunt, with its 500 ohms resistance as against the few ohms of the shunted quarter-mile, divert one-hundredth part of this minute fraction from the District wire? Plainly, the music reproduced in the Providence telephone did not require one ten-thousandth, nor one hundred-thousandth, of the force originally imparted to the Albany wire.

In December 1877 Prof. E. Sacher of Vienna undertook some careful investigations with a view of measuring the inductive effect in telephone circuits. He found that signals from three Smee cells sent through one wire, 120 metres long, could be distinctly heard in the telephone on another and parallel wire 20 metres distant from it.[1]

Early in 1879 M. Henri Dufour tried similar experiments, and with the same results. Two covered copper wires were stretched parallel over a length of 15 metres, and at distances apart varying from 15 to 45 centimetres. In connection with one of the wires were the battery and the ordinary Morse apparatus, the gas-pipes being used to complete the circuit. The ends of the other wire were joined to the telephone so as to form a complete metallic circuit. The current employed produced a deflection of 60° on the galvanometer. Under these conditions all the motions of the key were distinctly heard in the telephone, and the author was satisfied that a telegraphist would have understood the signals, even when the distance between the two wires was 45 centimetres.[2]

When we consider the shortness of these wires, the effects are sufficiently striking; but before this, equally

[1] 'Electrician,' vol. i. p. 194. [2] Ibid., vol. ii. p. 182.

striking results had been obtained on actual telegraph lines, where there was no battery, and where the infinitesimal currents produced by speaking into a Bell telephone on one wire were able to induce currents in a parallel wire sufficient to render the words audible in another telephone in its circuit. Dr Channing found this to be possible "under very favourable conditions."[1]

Another striking illustration is furnished by Prof. Blake, of Brown University, U.S., who talked with a friend for some distance along a railway (using the two lines of rails for the telephonic circuit), hearing at the same time the Morse signals passing along the telegraph wires overhead.[2]

PROFESSOR JOHN TROWBRIDGE—1880.

Such are a few of the early instances noted of the extreme sensitiveness of the telephone, by the aid of which the problem of wireless telegraphy was now to be attacked with a fair measure of success, and advanced a long way towards a practical solution.

Mr J. Gott, then superintendent of the Anglo-American Telegraph Company at St Pierre, was, I believe, the first to suggest the employment of the telephone in this connection. In a brief communication, published in the 'Jour. Inst. Elec. Engs.' (vol. vi. p. 523), he says: "The island of

[1] For a curiously similar case, the result of a wrong connection of the line wires, see the 'Telegraphic Journal,' vol. ix. p. 68.

[2] The absence of insulation in this experiment recalls the fact that a telephone line using the earth for the return circuit often works better when the insulation is defective, as it is then less affected by extraneous currents. Thus, in 1882, the Evansville (Ind.) Telephone Exchange Company worked 400 miles of line without insulators of any kind (the wires being simply attached to the poles), and generally with better results than when insulators were used. ('Electrician,' vol. ix. p. 481.)

St Pierre is, perhaps, better insulated than most places. Hundreds of yards from the station, if a wire be connected to earth, run some distance, and put to earth again, with a telephone in circuit, the signals passing through the cables can be heard."

There are two offices on the island,—one used for repeating the cable business on the short cables between Sydney, C.B., and Placentia, N.F., and operated by the Morse system, with a comparatively powerful battery; the other is the office at which the Brest and Duxbury cables terminate, and is furnished with very delicate instruments—the Brest cable, which is upwards of 2500 miles long, being operated by Thomson's exceedingly sensitive dead-beat mirror galvanometer; whilst on the Duxbury cable the same inventor's instrument, the siphon recorder, is used. The Brest instrument was found seriously affected by earth-currents, which flowed in and out of the cable, interfering very much with the *true* currents or signals, and rendering it a difficult task for the operator to decipher them accurately. The phenomenon is not an uncommon one; and the cause being attributed to the *ground* used at the office, a spare insulated wire, laid across the island, a distance of nearly three miles, and a metal plate connected to it and placed in the sea, was used in lieu of the *office ground*. This had a good effect, but it was now found that part of the supposed earth-currents had been due to the signals sent by the Morse operator into his wire, for when the recorder was put in circuit between the ground at the cable office and the sea ground—three miles distant—the messages sent by the Morse were clearly indicated,—so clearly, in fact, that they were automatically recorded on the tape.

It must be clearly understood that the two offices were in no way connected, nor were they within some 200 yards of each other; and yet messages sent at one office were distinctly read at the other, the only connection between the two being through the earth, and it is quite evident that they could be so read simultaneously at *many* offices in the same neighbourhood. The explanation is clear enough. The potential of the ground at the two offices is alternately raised and lowered by the Morse battery. The potential of the sea remains almost, if not wholly, unaffected by these, and the island thus acts like an immense Leyden jar, continually charged by the Morse battery and discharged, in part, through the short insulated line. Each time the Morse operator depressed his key he not only sent a current into his cable, but electrified the whole island, and this electrification was detected and indicated on the recorder.[1]

As the result of these experiences, Mr Gott gave it as his opinion that "speaking through considerable distances of earth without wires is certainly possible with Bell's telephone, with a battery and Morse signals."

Professor John Trowbridge of Harvard University, America, was, however, the first to systematically study the problem, and to revive the daring project of an Atlantic telegraph without connecting wires, and the less ambitious but equally useful project of intercommunication between ships at sea.[2] In fact, Trowbridge's researches may truly be

[1] See now Salvá's curious anticipation in 1795 of this phenomenon, p. 2, *ante*. The peculiarity, due to geological formation, is not confined to St Pierre; it is often met with in practice, though usually in lesser degrees. See some interesting cases, noted by G. K. Winter and James Graves, 'Jour. Inst. Elec. Engs.,' vol. i. p. 88, and vol. iv. p. 34.

[2] Mr H. C. Strong of Chicago, Illinois, claims to have suggested in 1857, in a Peoria, Ill., newspaper, the possibility of communication between ships at sea by means of a wireless telegraph then recently invented by his friend Henry Nelson of Galesburg. See Mr Strong's letter in the New York 'Journal of the Telegraph,' August 15, 1877.

said to form a new starting-point in the history of our subject, for, as we shall see later on, it is chiefly to him that Messrs Preece, Bell, and probably other experimenters in this field, owe their inspirations.[1] His investigations, therefore, deserve to be carefully followed.

The observatory at Harvard transmits time-signals from Cambridge to Boston, a distance of about four miles, and the regular recurrence of the beats of the clock afforded a good means of studying the spreading of the electric currents from the terminal of the battery which is grounded at the observatory. In all the telephone circuits between Boston and Cambridge, in the neighbourhood of the observatory line, the ticking of the clock could be heard. This ticking had been attributed to induction, but this, according to Prof. Trowbridge, is an erroneous conclusion, as he shows by a mathematical analysis into which we need not enter. The result goes to show that, with telephones of the resistance usually employed, no inductive effect will be perceived by the use of even ten quart Bunsen cells between wires running parallel, a foot apart, for a distance of 30 or 40 feet.

For this and other reasons, he says, it is impossible to hear telephonic messages by induction from one wire to another, unless the two run parallel and very close to each other for a long distance. This distance generally exceeds the limit at which the ordinary Bell telephone ceases to transmit articulate speech. The effects which have usually been attributed to induction are really, he says, due to the earth connections and to imperfect insulation.

Having determined in this manner that the echoes of the

[1] See pp. 92 and 137, *infra*. Professor Trowbridge's researches are given at length in a paper, "The Earth as a Conductor of Electricity," read before the American Academy of Arts and Sciences in 1880. See also 'Silliman's American Journal of Science,' August 1880, which I follow in the text.

time-signals observed on the telephone lines were not due to induction, but to leakage from the clock circuit, Prof. Trowbridge proceeded to study the extent of the equally electrified or equi-potential surfaces of the ground surrounding the clock battery. His method of exploration was to run a wire 500 or 600 feet long to earth at each end, including a telephone of 50 to 60 ohms resistance. Evidence of a current in this exploratory circuit was plainly shown by the ticking sound which making and breaking the circuit caused in the telephone, and the time-signals could be distinctly heard in a field 220 yards from the observatory where one earth of the time-signal wire is located. At a distance of a mile from the observatory, and not in the direct line between that place and the Boston telephone office, the time-signals were heard by connecting through a telephone the gas-pipes of one building with the water-pipes of another only 50 feet apart. In another experiment at the Fresh Pond lake in Cambridge, signals sent from Boston to Waltham (ten to twelve miles) were heard by simply dipping the terminal wires of the telephone in the lake, and some distance apart, where they must have been far away (? four miles) from the battery earth.

Prof. Trowbridge performed a large number of similar experiments, varied in every way, all going to prove (1) that a battery terminal discharging electricity to earth is the centre of waves of electrical energy, ever widening, and ever decreasing in strength or potential as they widen; and (2) that on tapping the earth in the way described at two points of different potentials (not very distant, if near the central source, and more removed the farther we recede from the source) we can obtain in the telephone evidence of their existence. Prof. Trowbridge then goes on to say:—

"In a discussion on the earth as a conductor, Steinheil

says: 'We cannot conjure up gnomes at will to convey our thoughts through the earth. Nature has prevented this. The spreading of the galvanic effect is proportional . . . to the square of the distance; so that, at the distance of 50 feet, only exceedingly small effects can be produced. . . . Had we means which could stand in the same relation to electricity that the eye stands to light, nothing would prevent our telegraphing through the earth without conducting wires.' [1]

"The telephone of Prof. Bell, though far from fulfilling the conditions required by Steinheil, is nevertheless our nearest approach to the desideratum.

"The theoretical possibility of telegraphing across the Atlantic without a cable is evident from the survey which I have undertaken. The practical possibility is another question. Powerful dynamo-electric machines could be placed at some point in Nova Scotia, having one end of their circuit grounded near them and the other end grounded in Florida, the connecting wire being of great conductivity and carefully insulated throughout. By exploring the coast of France, two points on surface lines not at the same potential could be found; and by means of a telephone of low resistance, Morse signals sent from Nova Scotia to Florida could be heard in France. Theoretically, this is possible; but practically, with the light of our present knowledge, the expenditure of energy on the dynamo-electric machines would be enormous." [2]

Professor Trowbridge has suggested the applicability of this method to the intercommunication of ships at sea.

[1] See p. 5, *ante*.

[2] A writer in the 'Electrician' (vol. v. p. 212), commenting on this passage, says: "Prof. Trowbridge seems to overlook the advantage of employing large condensers between the dynamo machines and the earth. They would prove of great service in exalting the earth potentials at the terminal stations."

Let, he says, a steamer be provided with a powerful dynamo. Connect one terminal of the dynamo with the water at the bow of the steamer, and the other to a long wire, insulated except at its extreme end, dragging over the stern, and buoyed so as not to sink. The current from the dynamo will thus pass into the water and spread out over a large area, as before explained, saturating, so to speak, the water with electricity. Suppose this current be interrupted by any suitable means, say one hundred times a second. Let the approaching steamer be provided with a telephone wire, the ends of which dip into the water at her bow and stern respectively. On entering the saturated area the telephone will respond to the interruptions of the dynamo by giving out a continuous buzzing sound. If now in the dynamo circuit we have a manipulating arrangement for breaking up the electric impulses into long and short periods, corresponding to the Morse alphabet, one ship can speak to the other. It is hardly necessary to add that by providing each steamer with a dynamo circuit and a telephone circuit reciprocal correspondence could be maintained, it being only necessary for the steamer desiring to listen to stop and disconnect the dynamo. The success of this method of communicating between ships in a fog depends upon the distance between the ends of the dynamo circuit and upon the strength of the current, or electrical impulses imparted to the water.

It is probable that a dynamo capable of maintaining one hundred incandescent lamps could establish a sufficient difference of potential between the water at the bow and at the end of a trailing wire, half a mile long, to affect a telephone on an approaching ship while yet half a mile distant.

In a discussion on Prof. Graham Bell's paper, read before the American Association for the Advancement of Science,

1884, Prof. Trowbridge described another plan, using instead of the telephone circuit a sensitive galvanometer connected up to a cross-arm of wire, whose ends dip into the water at each side of the ship. When one vessel comes within the area electrically saturated by another, the galvanometer will show how the equipotential lines are disturbed, and if a map of these lines be carefully traced we can fix the position of the approaching ship. He adds: "The method could also be applied to saturating the water around a rock, and you could take electrical soundings, so to speak, and ascertain your position from electrical maps carefully made out."

In a later paper published in the 'Scientific American Supplement,' February 21, 1891, Prof. Trowbridge discusses the phenomena of induction, electro-magnetic and static, as distinguished from leakage or earth conduction, and with reference to their employment in wireless telegraphy.

The hope, he says, that we shall be able to transmit messages through the air by electricity without the use of connecting wires is supposed by some to indicate its realisation at a future day. Let us examine how near we are at present to the realisation of this hope.

He supposes that the chief use of any method by which connecting wires could be dispensed with would be at sea in a fog. On land for considerable distances it is hardly probable that any electrical method could be devised in which air or the ether of space could advantageously replace a metallic conductor. The curvature of the earth would probably demand a system of frequent repetition, which is entirely obviated by the use of a wire. If, however, an electrical or magnetic system could be made to work through the air even at the distance of a mile, it would be of very great use at sea in averting collisions; for any system of signals depending upon the use of fog-horns or fog-whistles is apt to mislead on account of the reflection of the sound from layers of air of different densities and from the surface of the water. The difficulty of ascertaining the direction of a fog-horn in a thick fog is well known. The waves of sound, even if they are carefully directed by a trumpet or by parabolic reflectors, diverge so rapidly that there is no marked difference in the intensity between a position in the direct line and one far to one side.

The most obvious method of signalling by electricity through the air is by electro-magnetic induction. Suppose

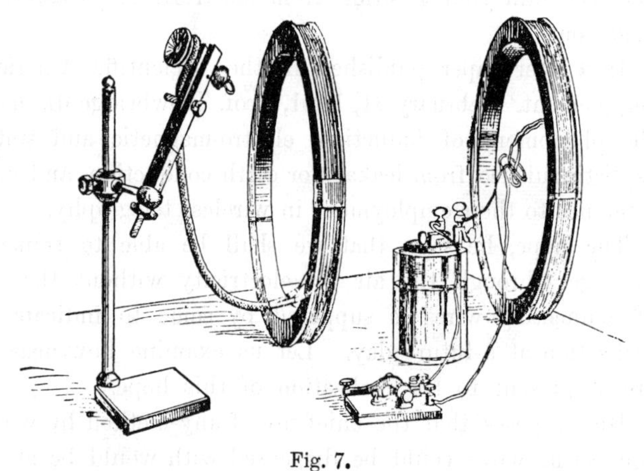

Fig. 7.

we have a coil of copper wire consisting of many convolutions, the ends of which are connected with a telephone (fig. 7). If we place a similar coil, the ends of which are connected to a battery through a key, within a few feet of the first and parallel to it, each time the current is made and broken in the battery coil instantaneous currents are produced by induction in the other coil, as can be heard by the clicks in the telephone.

To illustrate induction at a distance, Prof. Joseph Henry

placed a coil of wire, $5\frac{1}{2}$ feet in diameter, against a door, and at a distance of 7 feet another coil of 4 feet diameter. When contact was made and broken with a battery of eight cells in the first coil, shocks were felt when the terminal wires of the second were placed close together on the tongue.

In all such methods the wires or coils which produce an electrical disturbance in a neighbouring coil are never more than a few feet apart. Now let us suppose that a wire is stretched ten or twelve times, to and fro, from yard-arm to yard-arm of a steamer's foremast, and connected at the ends either with a powerful battery or dynamo, or with a telephone, as may be required either for signalling or for listening. Let an approaching steamer have a similar arrangement. If now the current on one vessel be interrupted a great number of times per second, a musical note will be heard in the telephone of the other vessel, and *vice versâ*. The sound will be strongest when the two coils are parallel to each other. If, therefore, the coils be movable the listener can soon find the position of greatest effect, and so fix the direction in which the signalling steamer is approaching.

It may not even be necessary to connect the telephone with the coil, for it has been found that if a telephone, pure and simple, be held to the ear and pointed towards a coil in which a current of electricity is rapidly interrupted, the makes and breaks will be heard, and this even when the wire coil of the telephone is removed, leaving only the iron core and the diaphragm.[1]

[1] Mr Willoughby Smith was, I believe, the first in recent times to observe these effects. See his paper on "Volta-Electric Induction," 'Jour. Inst. Elec. Engs.,' vol. xii. p. 457. But exactly similar effects, *mutatis mutandis*, were described by Page in 1837, to which he gave the name of Galvanic Music, and which he found to be due to the fact that iron when magnetised and demagnetised gave out a

Nothing could seem simpler than this, but, unfortunately, calculation shows that under the best conditions the size of the coils would have to be enormous. Prof. Trowbridge has computed that to produce an audible note in the telephone at a distance of half a mile, a coil of ten turns of 800 feet radius would be necessary; but it is evident that a coil of this size would be out of the question. Instead, however, of increasing the size of the coil beyond the practical limits of the masts and yard-arms, we could increase the strength of the current so as to be effective at the distance of half a mile; but, again, calculation shows that this strength of current would be beyond all practical limits of dynamo construction, unless we discover some method of tuning, so to speak, two coils so that the electrical oscillations set up in one may be able to evoke in the other sympathetic vibrations.[1]

Since, then, we have little, apparently, to hope for from electro-magnetic induction in signalling through a fog, cannot we expect something from static induction? This form of induction can be well illustrated by an early experiment of Prof. Henry. An ordinary electrical machine was placed in the third storey of his house, and a metal plate 4 feet in diameter was suspended from the prime conductor. On the first floor or basement, 30 feet below in a direct line, was placed a similar plate, well insulated. When the upper plate was charged by working the machine, the lower plate showed signs of electrification, as was evidenced by its effect on the pith-ball electroscope.[2]

sound. De la Rive, in 1843, rightly traced this sound to the slight elongation of iron under the magnetic strain—a fact which, in its turn, was first observed by Joule in 1842. For Page's discovery see the 'Magazine of Popular Science,' 1837, p. 237.

[1] Prof. Oliver Lodge is now engaged on this very problem. See 'Jour. Inst. Elec. Engs.,' No. 137, p. 799.

[2] See an excellent account of Henry and his work in the New

The distance to which this electrical influence can be extended depends upon the charging power of the machine and the dimensions of the plate. If we could erect an enormous metal plate on a hill, insulated and powerfully charged, it is probable that its electrical influence could be felt at the distance of the horizon; but here, again, the question of practical limits comes in as a bar, so that, at the present time (February 1891), this method of signalling without wires seems as little practicable as the others.

After following me in this study of Prof. Trowbridge, the reader may well begin to despair, for while the learned Professor's investigations are extremely interesting, his conclusions are very disappointing. But the darkest hour is just before the dawn, and so it is in this case.

PROFESSOR GRAHAM BELL—1882.

Following the lines suggested by Prof. Trowbridge, Prof. Bell carried out some successful experiments, an account of which is given in his paper read before the American Association for the Advancement of Science in 1884.

"A few years ago," he says, "I made a communication on the use of the telephone in tracing equipotential lines and surfaces. I will briefly give the chief points of the experiment, which was based on experiments made by Prof. Adams of King's College, London. Prof. Adams used a galvanometer instead of a telephone.

"In a vessel of water I placed a sheet of paper. At two points on that paper were fastened two ordinary sewing

York 'Electrical Engineer,' January 13, 1892, and succeeding numbers, from the pen of his daughter, Mary A. Henry. Abstracts of these papers are given in the 'Electrician,' vol. xxviii. pp. 327, 348, 407, 661.

needles, which were also connected with an interrupter that interrupted the circuit about one hundred times a second. Then I had two needles connected with a telephone: one needle I fastened on the paper in the water, and the moment I placed the other needle in the water I heard a musical sound from the telephone. By moving this needle around in the water, I would strike a place where there would be no sound heard. This would be where the electric tension was the same as in the needle; and by experimenting in the water you could trace out with perfect ease an equipotential line around one of the poles in the water.

"It struck me afterwards that this method, which is true on the small, is also true on the large scale, and that it might afford a solution of a method of communicating electrical signals between vessels at sea.

"I made some preliminary experiments in England, and succeeded in sending signals across the river Thames in this way. On one side were two metal plates placed at a distance from each other, and on the other two terminals connected with the telephone. A current was established in the telephone each time a current was established through the galvanic circuit on the opposite side, and if that current was rapidly interrupted you would get a musical tone.

"Urged by Prof. Trowbridge, I made some experiments which are of very great value and suggestiveness. The first was made on the Potomac river.

"I had two boats. In one boat we had a Leclanché battery of six elements and an interrupter for interrupting the current very rapidly. Over the bow of the boat we made water connection by a metallic plate, and behind the boat we trailed an insulated wire, with a float at the end carrying a metallic plate, so as to bring these two terminals about 100 feet apart. I then took another boat and sailed off. In this boat we had the same arrangement, but with a telephone in the circuit. In the first boat, which was moored, I kept a man making signals; and when my boat was near his I would hear those signals very well—a musical tone, something of this kind: tum, tum, tum. I then rowed my boat down the river, and at a distance of a mile and a quarter, which was the farthest distance I tried, I could still distinguish those signals.

"It is therefore perfectly practicable for steam-vessels with dynamo machines to know of each other's presence in a fog when they come, say, within a couple of miles of one another, or, perhaps, at a still greater distance. I tried the experiment a short time ago in salt water of about 20 fathoms in depth. I used then two sailing-boats, and did not get so great a distance as on the Potomac. The distance, which we estimated by the eye, seemed to be about half a mile; but on the Potomac we took the distance accurately on the shore."

Later, in urging a practical trial of his method, Prof. Bell further said: "Most of the passenger steamships have dynamo engines, and are electrically lighted. Suppose, for instance, one of them should trail a wire a mile long, or any length, which is connected with the dynamo engine and electrically charged. The wire would practically have a ground connection by trailing in the water. Suppose you attach a telephone to the end on board. Then your dynamo or telephone end would be positive, and the other end of the wire trailing behind would be negative. All of the water about the ship will be positive within a circle whose radius is one-half of the length of the wire. All of the water about the trailing end will be negative within a circle whose radius is the other half of the wire. If your wire is one mile long, there is then a large area of water about the ship which is affected either positively or negatively by the dynamo engine and the electrically charged wire. It will be impossible for any ship or object to approach within the water so charged in relation to your ship without the telephone telling the whole story to the listening ear. Now, if a ship coming in this area also has a similar apparatus, the two vessels can communicate with each other by their telephones. If they are enveloped in a fog, they can keep out of each other's way. The ship having the telephone can detect other ships in its track, and keep out of the way in a fog or storm. The matter is so simple that I hope our ocean steamships will experiment with it."[1]

PROFESSOR A. E. DOLBEAR—1882.

Prof. Dolbear of Tuft's College, Boston, was also, about the same time as Graham Bell, engaged on the problem of a wireless telegraph, and produced a very simple and workable apparatus, which he patented in the United States (March 1882), and of which he gave a description at a meeting of the American Association for the Advancement of Science in the following August. I take the following account from his specification as published in the 'Scientific American Supplement,' December 11, 1886:—

"In the diagram, A represents one place (say Tuft's College) and B a distant place (say my residence).

"C is a wire leading into the ground at A, and D a wire leading into the ground at B.

"G is an induction coil, having in the primary circuit a microphone transmitter T, and a battery *f'*, which has a number of cells sufficient to establish in the wire C, which is connected with one terminal of the secondary coil, an electro-motive force of, say, 100 volts. The battery is so

[1] 'Public Opinion,' January 31, 1886.

connected that it not only furnishes the current for the primary circuit, but also charges or electrifies the secondary coil and its terminals C and H′.[1]

"Now, if words be spoken in proximity to transmitter T, the vibration of its diaphragm will disturb the electric condition of the coil G, and thereby vary the potential of the ground at A, and the variations of the potential at A will cause corresponding variations of the potential of the ground at B, and the receiver R will reproduce the words spoken in proximity to the transmitter, as if the wires C D were in contact, or connected by a third wire.

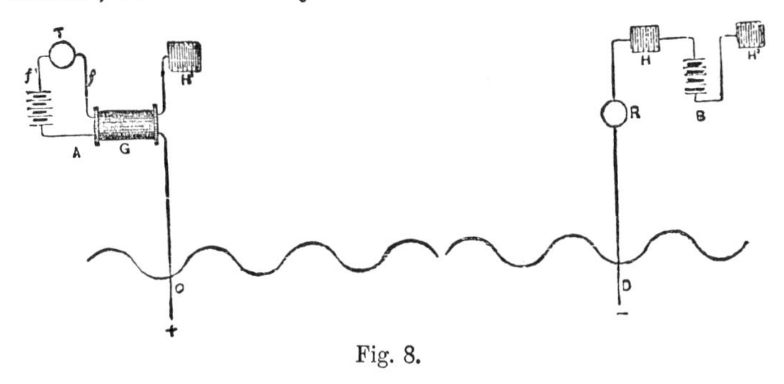

Fig. 8.

"There are various well-known ways of electrifying the wire C to a positive potential far in excess of 100 volts, and the wire D to a negative potential far in excess of 100 volts.

"In the diagram, H H′ H^2 represent condensers, the condenser H′ being properly charged to give the desired effect. The condensers H and H^2 are not essential, but are of some benefit; nor is the condenser H′ essential when the secondary coil is otherwise charged. I prefer to charge all these condensers, as it is of prime importance to keep the grounds of wires C and D oppositely electrified, and while, as is

[1] The diagram, which we have carefully copied, does not show how this is done, but the practical reader will easily supply the necessary connections.

obvious, this may be done by either the batteries or the condensers, I prefer to use both."

In the article from which I am quoting the author gives some additional particulars which are worth repeating. "My first results," he says, "were obtained with a large magneto-electric machine with one terminal grounded through a Morse key, the other terminal out in free air and only a foot or two long; the receiver having one terminal grounded, the other held in the hand while the body was insulated, the distance between grounds being about 60 feet. Afterward, much louder and better effects were obtained by using an induction coil having an automatic break and with a Morse key in the primary circuit, one terminal of the secondary grounded, the other in free air, or in a condenser of considerable capacity, the latter having an air discharge of fine points at its opposite terminal. At times I have employed a gilt kite carrying a fine wire from the secondary coil. The discharges then are apparently nearly as strong as if there was an ordinary circuit.

"The idea is to cause a series of electrical discharges into the earth at a given place without discharging into the earth the other terminal of the battery or induction coil—a feat which I have been told so many, many times was impossible, but which certainly can be done. An induction coil isn't amenable to Ohm's law always! Suppose that at one place there be apparatus for discharging the *positive* pole of the induction coil into the ground, say 100 times per second, then the ground will be raised to a certain potential 100 times per second. At another point let a similar apparatus discharge the *negative* pole 100 times per second; then between these two places there will be a greater difference of potential than in other directions, and a series of earth-currents, 100 per second, will flow from the one to the other. Any sensitive electrical device, a galvanometer or telephone, will be disturbed at the latter station by these currents, and any intermittence of them, as can be brought about by a Morse key in the first place, will be seen or heard in the second place. The stronger the discharges that can be thus produced, the stronger will the earth-currents be of course, and an insulated tin roof is an excellent terminal for such a purpose. I have generally used my static telephone receiver in my experiments, though the magneto will answer.

"I am still at work upon this method of communication, to perfect it. I shall soon know better its limits on both land and water than I do now. It is adapted to telegraphing between vessels at sea.

"Some very interesting results were obtained when the static receiver with one terminal was employed. A person standing upon the ground at a distance from the discharging point could hear nothing; but very little, standing upon ordinary stones, as granite blocks or steps; but standing on asphalt concrete, the sounds were loud enough to hear with the telephone at some distance from the ear. By grounding the one terminal of the induction coil to the gas or water pipes, leaving the other end free, telegraph signals can be heard in any part of a big building and its neighbourhood without *any connection whatever*, provided the person be well insulated."

When we come to speak of the Marconi system, we shall see how near Dolbear got to that discovery, or perhaps I should say how nearly he anticipated it. Comparing the arrangement, fig. 8 (especially when, as stated, a Morse key and automatic interrupter were used in place of the microphonic transmitter), with Marconi's, fig. 40, it will be seen that they are practically identical in principle. Dolbear's acute observation of the heightened effects obtained by projecting into free air the ungrounded terminals of the sending and receiving apparatus is his own discovery; while his use of condensers (answering to Marconi's capacity areas) and gilt kites carrying fine wire was another step in the right direction. Of course he does not use the Branly receiver, or the Righi sparking arrangement shown in fig. 40 (they were not known in 1882), but as regards the latter Marconi has himself discarded it in recent times, using a single spark-gap, which even is not absolutely necessary for the production of waves, leaving the secondary coil "open" alone sufficing.[1]

Prof. Dolbear's account of the action of his apparatus is in places a little puzzling, which, perhaps, can hardly be wondered at, for Hertz had not yet come to make clear the way which the American professor saw but as in a glass darkly. There can, however, be little doubt that he was using very long electric waves in 1882 (that is, five or six years before Hertz), and in much the same way as Marconi does now. When, for instance, he whistled into his microphonic transmitter, making it vibrate say 4000 times per second, did he not in effect start electric (now called Hertzian) waves $\frac{186000}{4000} = 46\frac{1}{2}$ miles long? We can easily see this now, but in 1882 the results were not so well understood. Dolbear was inclined to attribute them to some kind of ether action, obscure cases of which were then cropping up and attracting attention in the electrical world.[2]

Others thought that the results were "only extraordinary cases of electro-static induction." Thus Prof. Houston, who saw some of Dolbear's experiments and had himself repeated them, says: "The explanation of the phenomenon as I understand it would appear to be this—One of the plates of the receiver (that is, of the electro-static telephone) being connected through the body of the experimenter to the ground, partakes of the ground potential, while the other

[1] Broca, 'La Télégraphie sans Fils,' p. 89.

[2] See, for example, 'Telegraphic Journal,' February 15, 1876, p. 61, on The "Etheric" Force.

plate is *en rapport* with the free end of the sending apparatus by a line of polarised air particles. The experiment is simply an exceptional application of the principles of electro-static induction, and I am not at all sure that it is not susceptible of a great increase in delicacy, in which case it would become of considerable commercial value."[1]

Prof. Dolbear's friends in America are now claiming for him the discovery of the art of wireless telegraphy *à la Marconi.* They argue that Marconi arranges and works his circuits in the way substantially shown in Dolbear's patent of 1882; that he employs Dolbear's transmitting devices (induction coil, battery, and Morse key), as well as his aerial and ground connections on the sending and receiving apparatus. Dolbear emitted electric waves of many miles long, and received them on his electro-static telephone; Marconi, by using the same means, emits waves of many feet long, and receives them on a Branly coherer. Where, they ask, is the difference? Marconi's receiver is admitted to greatly extend the signalling range, but this does not affect the principle of the art, only its practical value, as to which they recall the fact that Graham Bell's telephone, as patented in 1876, was *practically* inoperative, yet the patent secured to him the honour and profit of the invention, as it was held that the principle was there, though in an imperfect form. All this is true, and I hope that Dolbear's early and for the time extraordinary experiments will always be remembered to his credit, but this, I think, should be done without detracting from the merit due to Marconi for his successful and, as I believe, entirely independent application of the same principle. But of this more anon.

[1] 'Scientific American Supplement,' December 6, 1884. At first, Dolbear's estimate of distance was modest—"half a mile at least," but it is said that recently he has worked his apparatus up to a distance of thirteen miles.

T. A. EDISON—1885.

Electric communication with trains in motion, like communication with ships at sea and with lighthouses, has long been a favourite problem with electrical engineers: indeed it is much the older of the two, and dates back to the first days of electric telegraphy.

In 1838 Edward Davy, the rival of Cooke and Wheatstone, proposed such a system. In a lecture on "Electric Telegraphy," delivered in London during the summer of 1838, he says:—

"I have a few words to say upon another application of electricity—namely, the purposes it will answer upon a railway, for giving notices of trains, of accidents, and stoppages. The numerous accidents which have occurred on railways seem to call for some remedy of the kind; and when future improvements shall have augmented the speed of travelling to a velocity which cannot at present be deemed safe, then every aid which science can afford must be called in to promote this object. Now, there is a contrivance, secured by patent, by which, at every station along the railway line, it may be seen by mere inspection of a dial what is the exact situation of the engines running either towards or from the station, and at what speed they are travelling. Every time the engine passes a milestone, the pointer on the dial moves forward to the next figure, a sound or alarm accompanying each movement.

"Not only this, but if two engines are approaching each other, by any casualty, on the same rails, then, at a distance of a mile or two, a timely notice can be given in each engine by a sound or alarm, from which the engineer would be apprised to slacken the speed; or, if the engineer be asleep or intoxicated, the same action might turn off the steam, independently of his attention, and thus prevent an accident."[1]

In 1842 William Fothergill Cooke published his 'Telegraphic Railways,' descriptive of a crude system of train signals, which was tried, in 1843, in the Queen Street tunnel, Glasgow, and in the Clay Cross tunnel, Derby; and, on a more extensive scale, in 1844, on the Great Eastern Railway, between Norwich and Yarmouth.

Dujardin in 1845, Brett and Little in 1847, Edwin Clark in 1854, Bonelli in 1855, and many others, proposed various systems of train signalling; but as they are all based on ordinary telegraphic principles and require connecting wires, they do not specially concern us in this history.

Mr A. C. Brown, an officer of the Eastern Telegraph Company, claims to have been the first to suggest, in 1881, the method of induction for communicating with moving trains. In a letter published in the 'Electrician,' March 21, 1885, he says:—

"My object was chiefly to provide an efficient means of fog-signalling, by enabling the signalman to communicate directly with the drivers or guards. I proposed to run a wire along the permanent way, parallel with the rails, and to wind a coil of wire round the engine, or carriage to be communicated with, in such a way as to get as long a length of wire parallel to, and as near to, the line-wire as possible, so as to be well exposed to the inductive action thereof. I then proposed to place in the signal-boxes a battery, signalling key, and rapid make-and-break instrument, or *buzzer*, and to thereby signal to the train, using a telephone in circuit with the train-coil as a receiver. By using an ordin-

[1] See the writer's 'History of Electric Telegraphy,' p. 407. The most perfect block system of the present day does not do anything like this. Davy's plan was actually patented by Henry Pinkus! See his patent specification, No. 8644, of September 24, 1840.

ary carbon transmitter in the line-wire, I also found it quite practicable to speak verbally to the train, so as to be distinctly heard in the telephone.

"This design was embodied in a paper which, in the year 1881, I laid before the managing director of the United Telephone Company, but want of time and opportunity prevented its being put into practice. It was experimentally tried at that time, using wire coils, properly proportioned in length, resistance, and distance apart to the conditions that would be obtained in practice. It has since been simplified and arranged to produce both visible and audible signals on the engine or car by induction from a No. 8 iron line-wire across a space of 6 inches, with a current of only one quarter ampère, or such as can easily be produced by the ordinary Daniell batteries used in railway work."[1]

In 1883 Mr Willoughby Smith threw out a similar suggestion towards the end of his paper on "Voltaic-Electric Induction," read before the Institution of Electrical Engineers, November 8 of that year:[2]—

"Telegraph engineers," he says, "have done much towards accomplishing the successful working of our present railway system, but still there is much scope for improvements in the signalling arrangements. In foggy weather the system now adopted is comparatively useless, and recourse has to be had at such times to the dangerous and somewhat clumsy method of signalling by means of detonating charges placed upon the rails.

"Now, it has occurred to me that Volta-Electric induction might be employed with advantage in various ways for signalling purposes. For example, one or more spirals could be fixed between the rails at any convenient distance

[1] For another proposal of Mr Brown, see p. 175, *infra.*

[2] Compare also his remarks, 'Jour. Inst. Elec. Engs.,' March 23, 1882, p. 144.

from the signalling station, so that, when necessary, intermittent currents could be sent through the spirals; and another spiral could be fixed beneath the engine, or guard's van, and connected to one or more telephones placed near those in charge of the train. Then, as the train passed over the fixed spiral, the sound given out by the transmitter would be loudly reproduced by the telephone, and indicate by its character the signal intended.

"One of my experiments in this direction will perhaps better illustrate my meaning. The large spiral was connected in circuit with twelve Leclanché cells and the two make-and-break transmitters before described. They were so connected that either transmitter could be switched into circuit when required, and this I considered the signalling station. The small spiral was so arranged that it passed in front of the large one at the distance of 8 inches, and at a speed of twenty-eight miles per hour. The terminals of the small spiral were connected to a telephone fixed in a distant room, the result being that the sound produced from either transmitter could be clearly heard and recognised every time the spirals passed each other. With a knowledge of this fact I think it will be readily understood how a cheap and efficient adjunct to the present system of railway signalling could be obtained by such means as I have ventured to bring to your notice this evening."

In 1885 Mr T. A. Edison had his attention directed to the subject, and with his usual thoroughness he soon produced a very complete system, with the assistance of Messrs Gilliland, Phelps, and W. Smith — to the last-named of whom the original idea is said to be due.[1]

[1] Although I have not seen any acknowledgment of their indebtedness, Mr Edison and his coadjutors can hardly have been ignorant of Mr Willoughby Smith's very clear proposal, of which their contrivance is but the practical realisation. Given the idea, the rest was easy enough.

The inevitable *avant-coureur* appeared in the technical journals of the period, and as it is delightfully characteristic of the great magician of Menlo Park, we venture to reproduce it here: "Mr Edison's latest invention, an arrangement to telegraph from moving trains, is thus described by a recent visitor to his laboratory: Overhead was a board eight inches wide, suspended from the ceiling by ropes fastened to one of its edges. One side of it was covered with tinfoil, and was facing toward a wall 20 feet distant. 'That,' said Mr Edison, 'is my railroad signal; I make electricity jump 35 feet, and carry a message. This is something quite new; no induction has ever been known that extended over 3 or 4 or 5 feet. This invention uses what is called static electricity, and it makes every running train of cars a telegraph station, accessible to every other telegraph station on the road. Messages may be sent to and from conductors, and to and from passengers. It requires no extra wire, either under the cars or at the side of the cars, but uses the ordinary telegraph just as it is put up at the side of the track. This white board is a receiver and transmitter. A board like it is to be fastened lengthwise along the peak of each car, where it will be out of the way and will not be a blemish. When the train is telegraphed to, the message jumps from the wire on the side of the track and alights on this board, and is conveyed to the apparatus in the train below. It works beautifully from those wires strung yonder. I was as much astonished as anybody at finding out what could be done. It costs very little, moreover, as 300 miles of road can be equipped for 1000 dols.'"

This contrivance was patented in England on June 22, 1885, in the joint names of T. A. Edison and E. T. Gilliland, and is fully described in their specification, No. 7583, of which the following is an abstract:—

The object of the invention is to produce apparatus for telegraphing between moving trains, or between trains and stations, by induction and without the use of connecting wires. The accompanying drawing (fig. 9) represents a station and portions of two trains with the apparatus for signalling. The carriage to be used as the signal office has placed upon its top or side, or upon each side, a metallic condensing surface running the entire length of the car. This consists of a strip *a* of metal, say a foot wide, well insulated by blocks of glass; or it may be thin sheet metal or metallic foil secured to canvas, and similarly insulated from the body of the car. To increase the total condensing surface, all the carriages of the train are preferably provided with such strips, which are connected electrically by suitable couplings *c* when the train is made up. A wire 1 is connected with this condensing surface, and

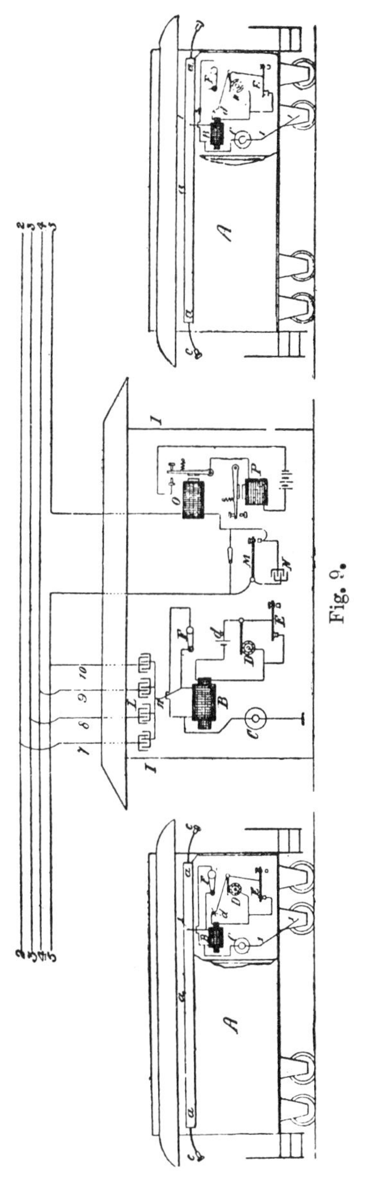

Fig. 9.

extends through the apparatus to the carriage-truck so as to form an earth connection through the wheels and the rails upon which they travel. The apparatus just mentioned consists of an induction coil B, the secondary wire of which is of extremely high resistance, and is in the circuit of wire 1, in which is also connected a telephone C of high resistance. This is preferably an electro-motograph telephone, the chalk cylinder of which is kept in constant rotation by a suitable motor, electrical or mechanical; but a magneto-electric or other suitable form of telephone may be employed.

In the primary circuit of the induction coil B are a local battery *d* and a revolving circuit-breaker D. This is a wheel having its surface broken by cross strips of insulation; upon it rests a spring, the circuit being through the spring to the spindle of the wheel. This wheel is kept in rapid motion by a suitable motor, electrical or mechanical, the current vibrations produced by it being a great number per second and audible in the telephone receiver.

The circuit-breaker is shunted by a back point key E, which, normally, short-circuits it and prevents it from affecting the induction coil. A switch F short-circuits the secondary wire of the induction coil when receiving, and is opened in transmitting.

The ordinary telegraph wires 2, 3, 4, 5, run on poles at the side of the track, and, grounded at their ends, are utilised collectively for conveying the signals. They form the other surface of the condenser (the strips on the carriages forming one surface), while the intervening body of air is the dielectric.

In signalling between trains, signals are transmitted by working the key E in the office upon one train. This causes static impulses at the condensing surface upon the carriages which affect the telegraph wires. These in turn affect the condensing surface upon the carriages of the other train, and cause impulses which are audible in the telephone.

At each signalling station I there is erected between the telegraph wires a large metallic condensing surface K (fig. 10). This may be attached to a frame supported from the

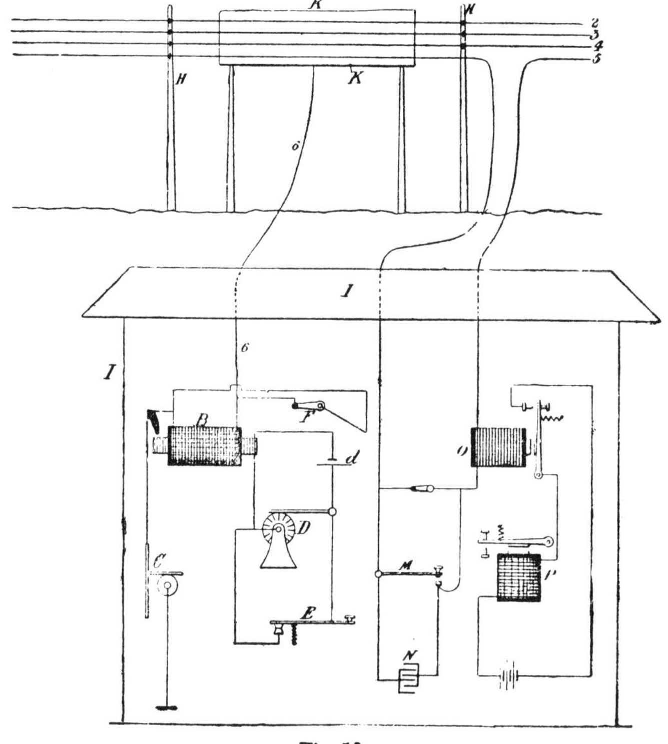

Fig. 10.

telegraph poles or from separate poles. A wire 6 runs from this condensing surface to the station, where it is connected to ground through the same character of transmitting and receiving apparatus already described for the carriages.

Instead of using this condensing surface outside of the

station, a separate wire (7, 8, 9, 10, fig. 9) may be attached to each telegraph wire (or to each of as many as it is desired to utilise) and run into the station, where it is connected to one side of a condenser L, of ordinary form. The other sides of the several condensers L are connected together, and by a common wire 11 to ground through the transmitting and receiving apparatus.

The telegraph wires are kept constantly closed for transmitting the induction impulses by shunting the regular Morse keys M by condensers N. These condensers do not interfere with the carrying on of the ordinary telegraphing over such wires, at the same time that they form constantly closed paths for the induction impulses independent of the working of the ordinary Morse keys. The ordinary Morse relay and sounder are shown at O and P respectively.

The stations being connected for railway signalling inductively with the line wires the same as are the trains, signals are received and transmitted by a station the same as by a train. The trains and stations are connected inductively with the line wires in multiple arc, so to speak, signals being transmitted by keys, circuit breakers, and induction coils, and received by telephones.

The signalling is conducted by Morse characters, or by numerical signals in accordance with an established code.

Speaking of the potentialities of his system, Edison, early in 1886, said: "The outcome is easy to predict. Special correspondents may, in the future, wire their despatches straight to the offices of their journals. Railway business will be expedited to a degree undreamt of as things are, and the risk of accidents will be largely diminished by knowing the position of trains and the cause of delay or accident, if any, at every stage of their route. Ships at sea, many miles apart, will be able to communicate by means of balloon-kites, soaring several hundred feet above their decks. Messages can be passed from ship to ship, and a casualty like that of the Oregon telegraphed to the nearest land. In times of war the applications of the air-telegraph system are obvious. Regions now remote from telegraphs could be brought within the civilised circle by means of mountain or forest stations equipped with the new apparatus. Even the man of business of the future may communicate with his employés as he journeys to and from his office, and save time or make money while he is literally on the wing. Not the least interesting feature of this new departure in telegraphy is the thought that, in its turn, it may be the harbinger of still more wondrous modifications of the system which has girdled the earth in a space inconceivably short when compared with that imagined by the fairy romancer who created Puck."[1]

The Edison system for trains was first put in operation at Staten Island, U.S.; then, a few months later, on the Chicago, Milwaukee, and St Paul line; and by October 1887 it was established on the Lehigh Valley Railroad, as related in the following paragraphs:—

"The success of what is called 'railway train telegraphy' is now assured, and October 6, 1887, will be a red-letter day in the history of the electric telegraph. On that day a special train left Jersey City with about 230 members of the Electric Club and guests of the Consolidated Railway Telegraph Company, in order to witness the working of the system on the Lehigh Valley Railroad. The system is a combination of the best features of the inventions of Edison, Gilliland, Phelps, and Smith, and although the speed often reached the rate of about sixty miles an hour, messages were sent from and received on the train without difficulty, although the current or the 'induction' had to jump from the train to the line wires, a distance of 25 feet. About

[1] 'Weekly Irish Times,' April 10, 1886.

four hundred messages were sent as the train ran from Perth Junction to Easton, amongst them a rather long one from Colonel Gouraud to Mr John Pender in London."[1]

"One of the most interesting triumphs of invention has been achieved on the Lehigh Valley Railroad during the snowstorms of the past winter in the United States. This railway for some months has been using on its trains the system of communication known as train telegraphy. The wire, being of steel, and stretched upon stout poles only 15 or 16 feet high, withstood the fury of the storm. The consequence was that all snowed-up trains on the Lehigh Valley Railroad kept up constant communication with the terminus of the road, could define exactly their position, and, in short, had all the advantages of perfect telegraphic communication."[2]

Soon after this the system fell into desuetude, and for a very simple reason—nobody wanted it. Whatever "special correspondents" and "the man of business" in the future may require, they, apparently, prefer nowadays to be free from telegrams of all sorts "while on the wing."

A few years later Mr Edison took out a fresh patent for the application of his method to long-distance communications over sea and land. The 'Illustrated London News' of February 27, 1892, gives an abstract of the specification with illustrative drawings.

If, says Mr Edison, a sufficient elevation be obtained to overcome the curvature of the earth, and to reduce as far as may be the earth's absorption, signalling may be carried on by static induction without the use of connecting wires. For signalling across oceans the method will be serviceable, while for communications between vessels at sea, or between vessels at sea and stations on land, the invention would be

[1] 'Public Opinion,' November 4, 1887.
[2] Ibid., April 13, 1888.

equally useful. There is also no obstacle to its employment between distant points on land, but in this case, he says, "it is necessary to increase the height (by using very high poles or captive balloons) from which the signalling operations are conducted, because of the induction-absorbing effect of houses, trees, and hills." These poles, surmounted by "condensing surfaces," are of course very like Marconi's —especially his earlier contrivances, where "capacity areas" in the shape of square sheets or cylinders of zinc are shown.

W. F. MELHUISH—1890.

We have seen (p. 39 *supra*) that the want of some form of wireless telegraph was peculiarly felt at a very early date in India, where the rivers are many and wide, and where for various reasons cables are liable to frequent breakage, causing interruptions which are as likely as not to be of long duration, owing to the great rush of waters and the flooding of banks.

I have already given some account of Dr O'Shaughnessy's experiments in this direction. It is all too short, but, unfortunately, it is all that I have been able to gather.

About the year 1858 Mr Blissett, a superintendent in the Indian Telegraph Department, resumed the inquiry, and obtained a fair measure of success by employing land-lines of considerable length on each bank of the river. In 1876 Mr Schwendler, then electrician, made some trials across the Hooghly at Barrackpore, near Calcutta, which were continued at intervals by his successor, the late Mr W. P. Johnston.

On September 9, 1879, this gentleman tried the following arrangement for signalling across the water of a canal. Fig. 11 shows the connections :—

E = 10 Bunsen's cells joined in series;

R, a needle instrument having a resistance of 1 ohm; also a telephone having a resistance of 4·25 ohms;

W = a resistance of 1 ohm
e = four Minotto cells joined parallel

} This arrangement exactly balanced the natural current through the receiving instrument.

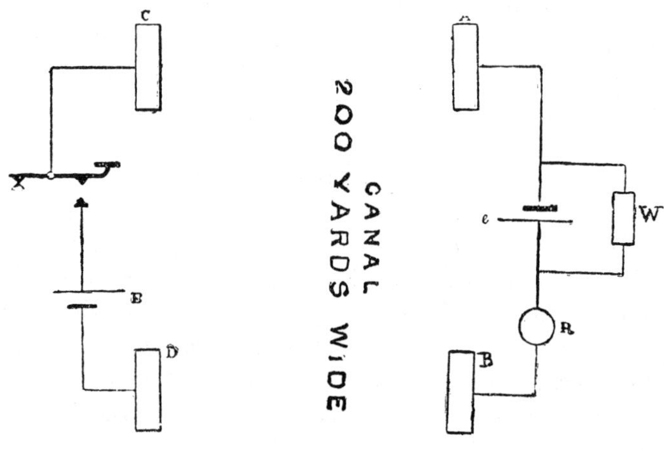

Fig. 11.

A, B, C, D were copper plates, 8 feet 8 inches by 4 feet 4 inches by 1-16th inch thick, buried on the banks of the canal. B was buried 15 yards distant from A, and D the same distance from C. All the plates were parallel to the canal. The resistance between A and B was 7·5 ohms, and that between C and D was the same. Under these conditions both the needle instrument and the telephone gave distinct and readable signals.

After several days of experiment with another method (fig. 12), using a single bare 600 lb. per mile galvanised wire, the following results were obtained:—

E = 15 Bunsen's cells in series;

R, a polarised Siemens relay of 21 ohms resistance;

$e = 4$ Minottos joined parallel } Balanced the natural
$w = 10$ ohms } current.

The signals received were quite regular and safe; the tongue of a relay worked an ordinary sounder in local circuit, and no difficulty was experienced in balancing the natural current through the relay.

A trial with bare wire for a distance of one and a half mile was not successful. Indeed, as it appeared that in order to obtain signals the battery power must be increased as the square of the distance, the limit of signalling through a bare wire under water is very soon reached.

Subsequently, three miles of the same wire, but partially

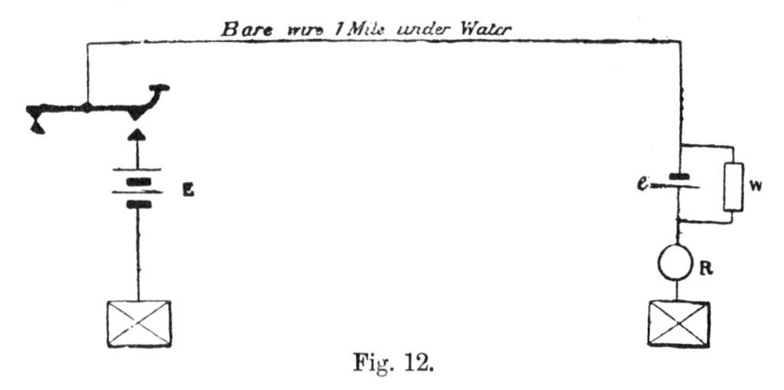

Fig. 12.

insulated by being passed through a mixture of pitch and tar, answered perfectly for the hour that the instruments were in circuit.

At various times during the year 1888 Mr Johnston carried out many experiments across canals and the river Hooghly, and as the result of these and other careful investigations he was led to the following conclusions:—

1. That up to one and a half mile it is perfectly easy to signal through a bare wire under water.
2. That for greater distances, judging from experiments, practical signalling is not possible.

In April 1889 Mr Johnston died, and the duties of electrician were entrusted to Mr Melhuish, who immediately took up the inquiry, and in the end produced some very considerable results, for which, I believe, the Government of India gave him the handsome honorarium of 5000 rupees.

The results of his investigations are embodied in a paper which was read before the Institution of Electrical Engineers on April 10, 1890. "Having studied," he says, "the recorded labours of my predecessor, and learnt that by pursuing the same lines it was hopeless to expect to be able to signal through a bare wire across a river that had a greater breadth than one and a half mile, I resolved to change the class of signalling apparatus and to continue the experiment. Discarding continuous steady currents and polarised receiving relays, I adopted Cardew's vibrating sounder, and the sequel will show how completely successful the change of instruments proved to be. I began from the beginning, and tried to signal across a water-way without a metallic conductor by laying down two earth-plates on each of its opposite banks. Readable signals having been exchanged, the distance separating each pair of plates was varied, with the view of ascertaining how close the plates might be brought together, the signals remaining still readable. Readable signals were exchanged when the distance separating the plates was equal to the breadth of the river, reading becoming more difficult as the plates were made to approach each other, and clearer and more distinct as the distance between the plates was made to exceed the breadth of the river. I learnt from these experiments that in order to obtain signals of sufficient distinctness for the practical purpose of transmitting messages, it would be necessary to construct a line on each bank of a river much longer than the breadth of the river; and as the rivers along the coasts in India are extremely wide, I became impressed with the impracticable character of such an undertaking, and decided to strike out a new line.

"This new line was the laying of two bare uninsulated iron wires across the water-way parallel to each other, and separated by a certain distance, the ends on each bank being looped together by means of an insulated conductor. Hence, though much of the circuit was laid under water, it was nevertheless a continuous metallic circuit. Beginning first with a complete square, by laying the wires as many yards apart as the river was wide, signals were instantly exchanged that were incomparably louder than those that were exchanged when the same area was bounded by four earth-plates. The length of each of the two wires under water was next gradually increased to 740 yards, and the distance separating them gradually diminished to 35 yards, the strength of the signals diminishing proportionately, and ceasing to be readable when the wires were further approached. The conclusion arrived at from these experiments was that, for the practical and useful purpose of signalling messages across a broad river, in the absence of an insulated cable, a complete metallic circuit was at least desirable. Acting on this conclusion, it was sought to apply it practically, and the following experiment was carried out: At a distance of fifteen miles west of Calcutta a cable is laid across the river Hooghly, which at this point is 900 yards wide. The iron guards of this cable were employed to form one of the metallic conductors, and at a distance of 450 yards down-stream a single wire, weighing 900 lb. per mile, was laid across the river to form the second metallic conductor, insulated land-lines having been run up to loop the two parallel conductors together. The experiment was quite a success, the signals being readable without difficulty.

"An experiment was next made on a defective cable across Channel Creek, at the mouth of the river Hooghly. This creek is crossed by two cables laid in the same trench; the length of each is 3000 yards, and one of them had been completely parted by a steamer's anchor. Several attempts were made to signal across by using the guards of one of the cables as a lead, and the guards of the other as a return wire, but the efforts proved unsuccessful owing to the too close proximity of the cables. For every crossing there is a certain minimum distance apart at which the cables must be laid, and if this minimum, which depends on the breadth of the river, be exceeded, an absolute short-circuit becomes established. But although it was not possible here to signal through the iron guards, the most perfect signals were passed through the two conductors when they were formed into a loop, notwithstanding the fact that the two ends of the broken conductor were exposed in the sea and were lying at a considerable distance apart. An experiment was now made in order to ascertain what chance there might be in the future of signalling across the two conductors, should an accident occur to the good cable. Accordingly, the conductor of the good cable was disconnected in the cable-house from the signalling apparatus and placed upon the ground, when the signals, though greatly diminished in volume, still continued to be distinctly readable. It may, therefore, be reasonably inferred that should the good cable suffer a similar fate to that of the defective cable, communication can, by means of Cardew's sounders, be kept up by looping the ruptured conductors until arrangements can be made for laying a new cable or repairing the defective ones.

"It will probably suffice if from the succeeding experiments that were made to test the efficiency of the vibrating sounder in the case of conductors breaking down at river crossings I select the following three, exhibiting as they do progressive evidence of the value of this signalling instrument, and culminating in establishing it beyond dispute as one that can be relied on for carrying on independent communication through the iron guards of cables while the insulated copper conductors form parts of other circuits.

"*Experiment No.* 1.—The local line from the Central Office, Calcutta, to Garden Reach is about four miles in length, and at about midway the wire spans a small river. Vibrating sounders having been put in circuit at each end of this line, the wire where it crosses the river was taken down and laid along the bed of the water-way. Signals were loud and clear at both ends.

"From the success of this experiment it may be inferred that on any ordinary line, should the wire from accidental causes come off the insulator and make earth by touching the bracket, standard, or ground, or should the wire break and both ends of it be lying on the ground or in a water-course, communication could still be maintained by means of the vibrating sounders.

"*Experiment No.* 2.—The line wire which connects the town of Chandernagore with Barrackpore is about ten and a half miles long, 900 yards of which consist of a cable laid across the river Hooghly. Vibrating sounders having been joined up in the telegraph offices at Barrackpore and Chandernagore, the insulated conductor of the cable was thrown out of circuit, and the line wire on each side of the river was joined to the iron guards of the cable. Thus for a length of half a mile out of ten and a half miles the conductor was wholly under water, yet it was found quite feasible to transmit messages between the two offices.

"From the success of this experiment it may be reasonably inferred that in the case of certain cable crossings, where the rivers are not too wide, should the copper conductor of the cable make dead earth, or become insulated by parting, communication could still be kept up between the two offices on either side.

"*Experiment No.* 3.—The terminus of the Northern Bengal State Railway at Sara is separated from that of the Eastern Bengal State Railway at Damukdia by the river Ganges. The opposite banks of the river in this locality are connected by two independent cable crossings. The length of one of these crossings is one mile 610 yards, and of the other four miles. The distance which separates the two cable-houses on the Damukdia side is three miles 1584 yards, and on the Sara side the cable-houses are only one mile 211 yards apart, giving a mean lateral distance in alignment of two miles 880 yards. The two cable-houses on each bank of the river have an insulated connecting land-line.

"The connecting land-lines having been joined to the iron guards of the cables, two vibrating sounders were placed in circuit, one on each side of the river, when signals so strong were transmitted across that it was not difficult to read them at a distance of 6 feet away from the receiving telephone.

"From the marked success of this experiment it may be inferred that at all river cable crossings where the cables are laid in separate alignments (and the farther apart the better), should the cables become interrupted, communication may still be maintained from bank to bank by using vibrating sounders, thus avoiding the delay, inconvenience, and cost of a boat service.

"It should also be remembered in the case of such a parallel cable crossing that, besides the circuits afforded by the copper conductors when these are in working order, there is always an additional local circuit available by means of the iron guards between the opposite cable-houses, and that this circuit could be used by means of the vibrating sounder as a talking circuit, in cases of necessity, without interrupting through working on either of the cables.

"It is desirable in circumstances similar to these to reduce all the resistance external to the actual connecting lines to as small a quantity as possible, and therefore, when messages are being transmitted, the telephone at the sending

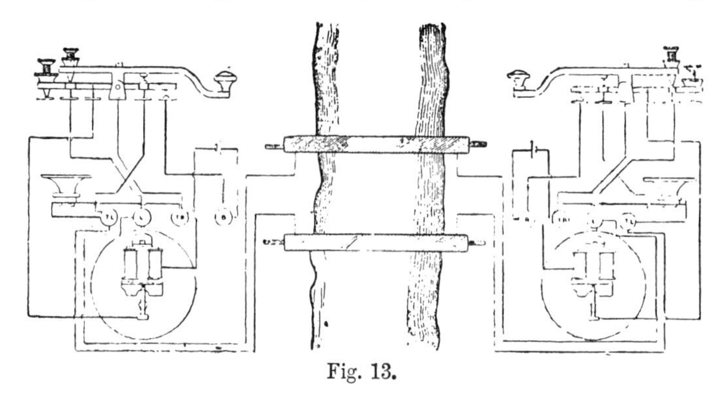

Fig. 13.

end should be removed from the circuit, as also should the vibrator from the receiving end. To effect this twofold purpose a special form of signalling key is requisite, and should be used. The action of this key, together with the complete set of connections for a parallel cable crossing, is shown in fig. 13." [1]

CHARLES A. STEVENSON—1892.

Early in 1892 Mr Charles A. Stevenson of the Northern Lighthouse Board, Edinburgh, threw out the suggestion that telegraphic communication could be established between ships at sea and between ship and shore by means of coils. [2]

[1] Melhuish's plan is the practical realisation of the early proposals of Highton and Dering. See pp. 40, 48, *supra*.

[2] 'Engineer,' March 24, 1892.

He tried many experiments in the course of that year, the results of which he reported to the Royal Society of Edinburgh on January 30, 1893. In this paper[1] he describes two methods of communicating between the shore and a ship, each of which supposes a cable to be submerged along the coast, and to be earthed in the sea—presumably (for the account is not clear) through an induction coil or transformer.

In the first method the ship has a wire, with a telephone in circuit, stretched from bow to stern, and terminating in coils which dip into the water, and which may or may not be insulated. When the ship approaches or crosses the cable at a right angle, or nearly so, the currents set up in the latter by a magneto-electric machine at the shore end are rendered audible in the telephone on board. If the coils be in the line of the cable, as they will be when the ship is over it lengthways, or approaches it broadside on, no sound is heard in the telephone, thus indicating the position of the vessel with respect to the known direction of the cable. An insulated wire, 400 feet long, was laid through a small lake (Isle of May) of brackish water, 15 feet deep. Alternations of current were set up in this wire by the bobbins of three-fifths of a De Meritens' magneto-electric machine (yielding 80 volts at its terminals). A small boat, having a wire with a telephone in circuit stretched from bow to stern and terminating in coils dipping into the water 10 feet apart, was rowed about in the vicinity of the submerged wire, and it was found that the currents in this wire were distinctly audible in the telephone up to a distance of over 300 feet.

The second method described by Mr Stevenson consisted in dropping into the sea from the deck of a ship a large electro-magnet (3 feet long, with 2000 turns of one-

[1] On Induction through Air and Water at Great Distances without the use of Parallel Wires.

eighth inch copper wire) with a telephone in circuit. The interruptions of the current from six dry cells through a wire 200 feet long could be heard in the telephone at a distance of 40 feet in air, while with twelve dry cells the effect was audible through 60 feet of salt water. Indeed, he says, there seemed to be little difference whether the medium was air, or fresh or salt water.

The first described method was practically tried in America, early in 1895, by Professor Lucien Blake, and was favourably spoken of in his report to the American Lighthouse Board, September 1895. A lightship is moored in 65 feet of water and four-and-a-half miles off Sandy Hook. Out from the shore an armoured cable was laid, terminating in a transformer, the core of the cable being earthed on the armour through the high-resistance coil of the transformer, while the terminals of the lower-resistance coil were earthed in the following manner. Three insulated copper wires, each one quarter mile in length, were laid parallel on the sea bottom and 300 feet apart. At one end they were connected together and joined to one terminal of the transformer, while the other and distant ends were earthed by means of pieces of wire netting about 20 feet square. The other terminal of the transformer was earthed by a similar piece of netting. At the centre of this "grid" arrangement the ship's moorings were fixed. The connections on board were as follows: The two hawse pipes were connected in the pipe by a copper bar, and extra plates were put between the metallic sheathing and the hawse pipes so as to ensure good sea connection. From the copper bar an insulated wire was carried to a telephone in the after-cabin, thence a wire from the other telephone terminal was carried aft and connected to a tail-piece of flexible conductor over the stern and dipping into the water.

When intermittent currents were sent into the cable from

the shore, there was set up in the area under the ship "an unequal electrical distribution, such that the potentials were of sufficient difference at the two ends of the ship to operate the telephone on board. Experiments showed that sufficient difference existed between bow and rudder sheathing, and even between bow and stern sheathing, to operate the telephone, but the effect was greatest with bow sheathing and stern tail-rope."

In another experiment on board the lighthouse tender Gardinia, the telephone circuit terminated in two plates, 7 feet by 3 feet, submerged from bow and stern, a distance of 113 feet. Here too, "sufficient difference of potential existed between the plates to make conversation with the shore possible while the tender was steaming about in the neighbourhood."

Mr Stevenson speaks of this as an electro-static effect, but as I understand it, and certainly as it has been tried by Prof. Blake, the method seems to belong more to the conductive order, and to be identical with that of Messrs Smith & Granville, to be presently described (p. 165, *infra*).

Mr Stevenson calls his second method "electro-magnetic," in contradistinction to the first or "electro-static" one, and with certain dispositions of the submerged cable it *might* be available for communicating between the shore and a lightship through a few fathoms of water. It is, however, interesting to us as being a step forward in the evolution of Mr Stevenson's ideas from conductive to inductive methods.

In a further paper, read before the Royal Society of Edinburgh, March 19, 1894, he describes his experiments with *insulated* coils of wire, or more correctly *spirals*, and says that a trial of his new method on a large scale had recently been made with a view of ultimately employing it for effecting communication between Muckle Flugga, in the Shetlands, and the mainland.

As regards the efficacy of the principle, the inductive effect of one spiral on another at a distance has long been known ; but hitherto, even with a very strong battery, it was impossible to bridge a greater distance than 100 yards, which for practical purposes was, of course, useless.

It is evident that if two coils are placed vertically so that their axes are coincident, their planes being parallel, or if they be placed so that their planes are in the same plane, they will be in good positions for electric currents sent in one to be apparent by induction in the other. For a small diameter, and where the electrical energy is small, the first position is suitable ; but where the energy is great and the diameter of coils great—in fact, when it is wished to carry the induction to many times the diameter of the coils—then it will be found that it is better to let the two coils be in the same plane, as it becomes impracticable to erect coils of large diameter with their planes vertical, but it is easy to lay them on their sides.

Mr Stevenson made a large number of laboratory experiments on the interaction of coils, with the view of calculating the number of wires, the diameter of coils, the number of ampères, and the resistance of the coils that would be necessary to communicate with Muckle Flugga ; and, after a careful investigation, it was evident the gap of 800 yards could, with certainty, be bridged by a current of one ampère with nine turns of post-office wire in each coil, the coils being 200 yards in diameter, and with two good telephones on the hearing coil.

Two coils, on telegraph-poles and insulators, were erected at Murrayfield, one coil being on the farm of Damhead and the other on the farm at Saughton, and as nearly as was possible on a similar scale, and the coils of similar shape to what was wished at Muckle Flugga. On erecting the coils, communication was found impossible, owing to the induction currents from the lines from Edinburgh to Glasgow, the messages in those lines being quite easily read, although the coils were entirely insulated and were not earthed. The phonophore which the North British Railway Company have on their lines kept up nearly a constant musical sound, which entirely prevented observations. On getting the phonophore stopped, it was found that 100 dry cells, with 1·2 ohms resistance each and 1·4 volts, gave good results, the observations being read with great ease in the secondary by means of two telephones. The cells were reduced in number down to fifteen, and messages could still easily be sent, the resistance of the primary being 24 ohms and the secondary no less than 260 ohms. If the circuit had been of good iron, with soldered joints and well earthed, the resistance would have been only 60 ohms. The induced current generated in the secondary would therefore be in the ratio of 480 [? 520] to 210; or, allowing for the resistance in the two telephones, we get practically only half the current we would have got if the line had been a permanent in place of a temporary one.

A trial was made of the parallel-wire system:[1] with 20 cells the sound was not heard, and with 100 cells it was heard as a mere scratch in comparison with the sound with the coil system with only 15 cells. A trial was made with the phonophore: the coils worked with 10 cells with perfect ease, and a message was received with only 5 cells. Speech by means of Deckert's transmitter was just possible, but it is believed that if the hearing circuit had been of less resistance it would have been easy to hear.

"It is difficult," says Mr Stevenson, "to understand how this system of coils, in opposition to the parallel-wire system, has not been recognised as the best; for assume that, with the arrangement we had, we heard equally with 100 cells by

[1] *I.e.*, Preece's method, to be presently described. See p. 144 *et seq.*, *infra*.

both systems, both having the same base (200 yards), then, by simply doubling the number of turns of wire on the primary and using thick wire, the effect would have been practically doubled, whereas by the parallel-wire system there is nothing for it but to increase the battery power. The difficulty of the current is thus removed by using a number of turns of wire. It must always be borne in mind that the effect is the result of simply increasing the diameter, keeping current and resistance the same. The larger the diameter the better. What is wanted is to get induction at a great distance from a certain given base with a small battery power, and the laboratory experiments and the trials in the field show that the way to overcome the difficulty of the current is by using a number of turns of wire. The secret of success is to apportion the resistance of primary and secondary, and the number of turns on each, to a practical battery power."

1. *Coil System.*—At 870 yards from centre to centre of coils, averaging each 200 yards diameter, with nine turns of wire, it was found that with a phonophore messages were sent with five dry cells, the resistance in primary being 30 ohms and the resistance of secondary 260 ohms, the current being 0·23 ampère, which, with nine turns, gives 2 ampère turns.

2. With a file as a make and break, it worked with 10 cells, giving 0·4 ampère or 3·6 ampère turns.

3. *Parallel-Wire System.*—With a file as a make and break, and with parallel lines earthed, it was heard with 100 cells, giving 1·1 ampère.

The primary coil circuit was entirely metallic in the Murrayfield trials, as it would have to be if erected at Muckle Flugga; but the secondary coil was earthed. When, however, the secondary was also made a complete insulated metallic circuit, with eight turns of wire, there seemed to be little difference in the result.

The calculation of the diameter necessary to hear at a

given distance is simple, from the fact that the hearing distance is proportional to the square root of the diameter of one of the coils, or directly as the diameter of the two

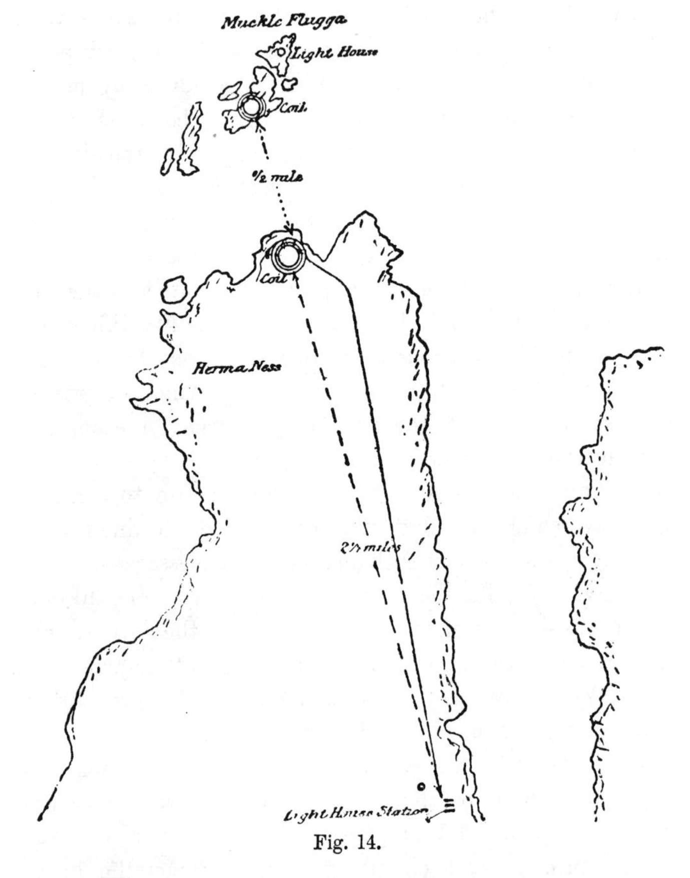

Fig. 14.

coils, so that, with any given number of ampères and number of turns, to hear double the distance requires double the diameter of coils, and so on.[1]

[1] Professor Lodge has recently shown that the law of distance is not the square root of diameter, but the two-thirds power, with a

In concluding his paper, Mr Stevenson says :—

"It has been attempted to be shown that the coil system is not only theoretically but practically the-best ; and I trust that we will soon hear of the Admiralty, &c., experimenting with it, and ultimately putting it in practice. Meantime my brother has recommended the Commissioners of Northern Lighthouses to erect the coil system at Muckle Flugga, and the Commissioners have approved ; and I hope soon to hear of the erection of this novel system of communication at the most northern point of the British Isles, as well as on our warships to assist in their manœuvring, by the establishment of instantaneous communication unaffected by wind or weather.

"The application of the coil system to communication with light vessels is obvious—viz., to moor the vessel in the ordinary way, and lay out from the shore a cable, and circle the area over which the lightship moorings will permit her to travel by a coil of the cable of the required diameter, which will be twice the length of her chain cable. On board the vessel there will be another coil of a number of turns of thick wire. Ten cells on the lightship and ten on the shore will be sufficient for the installation."[1]

given primary current ; and so doubling the circumference of each coil will permit signalling over more than double the distance, if other things can be kept the same. For such magnification, however, the thickness of the wire must be magnified likewise, or else more power will be consumed in the enlarged coil. 'Jour. Inst. Elec. Engs.,' No. 137, p. 803. Possibly Mr Stevenson did not take into account the increase in resistance owing to the increased length of wire, so that for practical purposes his formula may be sufficiently accurate.

[1] On May 28, 1892, Mr Sydney Evershed patented a similar method of using coils in connection with his very delicate receiving instrument or relay. The plan was actually tried in August 1896 on the North Sand Head (Goodwin) lightship. One extremity of the cable was coiled in a ring on the bottom of the sea, embracing the whole area over which the lightship swept while swinging to the tide, and

In a recent communication[1] Mr Stevenson gives some additional particulars. Referring to his proposed installation at the North Unst lighthouse, on Muckle Flugga, he tells us a gap of half a mile had to be bridged. The Commissioners of Northern Lighthouses, being impressed with the experiments shown them on a small scale—through stone and mortar—and on a larger scale at Murrayfield, decided on installing the system on Muckle Flugga; but, subsequently, financial difficulties arose, and the project was allowed to drop.

"It is well to remember," he says, "that in the Murrayfield trials a small number of cells was purposely used. Theory and formulæ give one the impression at first sight that a single outstretched wire is always best—the simple fact of getting a greater effect at a distance as a coiled wire is uncoiled and made straight supporting this impression; but formulæ, if they are to be practical, ought to take into account a *limited* area and *workable* amounts of resistance, current, &c., and then the fact is disclosed that the coiling of wires (whether condensers be used with them or not) becomes an advantage for most work which the engineer will be called upon to deal with.

"It is not necessary, as has been stated, that the coils should be identical in size and shape. Far from it; each case must be treated for size and configuration by itself.

the other end was connected with the shore. The ship was surrounded above the water-line with another coil. The two coils were separated by a mean distance of about 200 fathoms, but communication was found to be impracticable. The screening effect of the sea water and the effect of the iron hull of the ship absorbed practically all the energy of the currents in the coiled cable, and the effects on board, though perceptible, were very trifling—too minute for signalling. See Evershed's paper on Telegraphy by Magnetic Induction, Jour. Inst. Elec. Engs.,' No. 137, p. 852; also Stevenson on Telegraphy without Wires, 'Nature,' December 31, 1896.

[1] 'Jour. Inst. Elec. Engs.,' No. 137, p. 951; also No. 139, p. 307.

For instance, in the case of Muckle Flugga, my design was for a line two miles in length on the mainland, with a coil at the end enclosing a larger area than the one on the rock, which latter was opened out to the maximum possible. Again, in the case of Sule Skerry and the Flannan Islands, on the north-west of Scotland, where telegraphy by induction would be of great value, it would be impossible to make the coils of large diameter, but the coil on the mainland should be of large dimensions; indeed a single long wire with the ends earthed would be, perhaps, the best arrangement.

"For guarding a dangerous coast, a similar wire of many miles in length would be suitable for communicating warning signals to vessels on board of which were detectors, with coils necessarily of small dimensions. There are two ways of doing this, both of which I have tried. First, by means of a submarine cable along the line of coast. In this case the currents set up in the cable have to bridge only the sheet of water to the vessel, say twenty fathoms; or, if an electro-magnet be let down from the ship, only four or five fathoms. But here the cost and maintenance of a cable would be a weighty objection. The other way is to erect a pole line on shore, either along the coast or in the form of a coil on a peninsula. The main difference from the first plan is that the currents would have to be stronger to bridge the distance of several miles instead of a few fathoms; but the cost in comparison with a cable would be very small. I have tried this system with two miles of pole line and a coil about a quarter of a mile distant with perfect and never-failing success.

"I have made numerous trials of the coil *versus* parallel-wire system since 1891, and I have found—and other observers seem also to have found—that it is not practical to work the latter more than three or four times the length of base; whereas by coils I have found it possible to work many times their diameter. Thus in 1892, at the Isle of May lighthouse, I signalled to a distance 360 times the diameter of an electro-magnet coil with currents from a de Meritens' magneto-electric machine. Again, at Murrayfield, I signalled four times the base with five dry cells; and I have in Edinburgh a coil with iron core 17 inches diameter, which with one cell can easily signal through a space twenty-five times its diameter."

PROFESSOR ERICH RATHENAU—1894.

The last example of a wireless telegraph with which we have to deal in this part of our history is an arrangement devised by Prof. Rathenau of Berlin, with the assistance of Drs Rubens and W. Rathenau, and which was found to be practicable up to a distance of three miles in water.

Reports of the experiments of Messrs Preece, Stevenson, and others in England having appeared in the technical journals on the Continent, Prof. Rathenau, at the request of the Berlin Electrical Society, undertook to make a thorough investigation of the subject *de novo*.

After a careful study of the work of these electricians he felt convinced that the favourable results obtained in England, especially by Preece, were largely due to conduction. To verify this opinion he commenced a course of rigorous experimentation; and to prevent inductive effects entering into the calculation he decided to use ordinary battery currents, and in one direction only.

The outcome of the inquiry was published in an article which he contributed to the Berlin 'Elektrotechnische Zeitschrift,'[1] from which I make a few extracts. When a

[1] Abstract in 'Scientific American Supplement,' January 26, 1895, which I follow in the text.

current is sent through two electrodes immersed in a conducting liquid, the electrical equilibrium between these electrodes is not effected in a straight line, but in lines which spread out in the manner shown in fig. 15. Now, if we place in the liquid medium an independent conductor of electricity, it will attract or condense upon its surface a certain number of these lines, which can be utilised for the excitation of a properly constructed receiving apparatus. The distance at which these electrical effects can be produced is found to depend upon two factors—the available current strength and the distance between the electrodes.

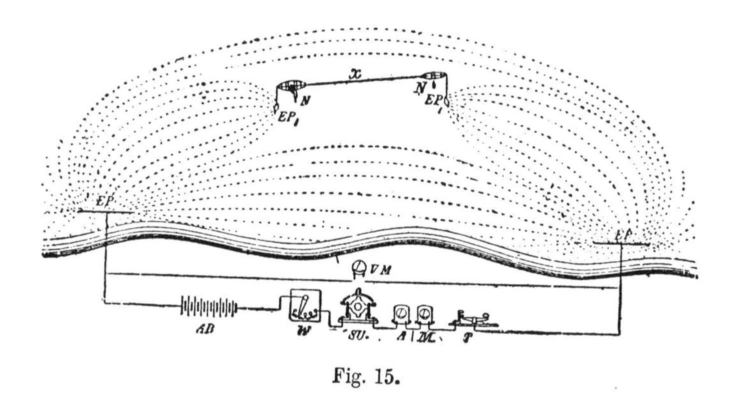

Fig. 15.

It was thought best to conduct the experiments on the lake Wannsee, near Potsdam, on account of the facilities in the way of apparatus afforded by the proximity of an electric-light station. The arrangement is shown in fig. 15. AB is a battery of 25 cells, W a set of resistance coils (0 to 24 ohms), SU an interrupter driven by a motor, AM an ampèremeter, VM a voltmeter, T a Morse key, EP EP two zinc plates immersed in the water, 500 yards apart, and connected by cable as shown. The receiving circuit comprises two zinc plates, EP_1 and EP_1, suspended by cable X from two boats, from

50 to 100 yards apart, and nearly three miles from the sending station; N N are telephones included in the circuit of X. For the purpose of transmitting signals, intermittent currents were sent from the battery, which, by depressing the key for long and short intervals, could be heard in the telephones as dashes and dots of the Morse code.

The object was to establish experimentally the best relation between the various factors—*i.e.*, the relation between the current strength in the primary circuit and the hearing distance for the telephones in the secondary circuit; the effect of various distances between the electrodes EP EP upon the clearness of the signals; the distance between EP_1 EP_1 which gave the most audible effect; and, finally, the effect of altering the shape and size of the plates.

On account of the non-arrival of some apparatus specially designed for these tests, the average current strength sent through the water did not exceed three ampères with 150 intermissions or current impulses per second. Again, the water of the Wannsee containing but a very small admixture of mineral salts offered a high resistance, so that it was found necessary to use large plates of 15 square yards surface.

With this arrangement no difficulty was encountered in the transmission of signals from the electric-light station to the boats anchored off the village of New Cladow—a distance, as has been said, of nearly three miles; and Prof. Rathenau was satisfied that, by a slight change in the construction of the ordinary telephone, signals could be sent over much greater distances.

"Lord Rayleigh," he says, "has stated that the sensitiveness of the telephone for currents with 600 reversals per second is about 600 times greater than for currents having but 130 reversals per second, but in my experiments the number of impulses did not exceed 150 per second. To get the best possible result in this system of transmission, a telephone should be used having a carefully tuned metallic tongue in place of the ordinary iron disc. Then, knowing the number of current-breaks in the primary circuit, the tongue should be so tuned as to vibrate in unison with that number, thereby producing much more distinct signals.

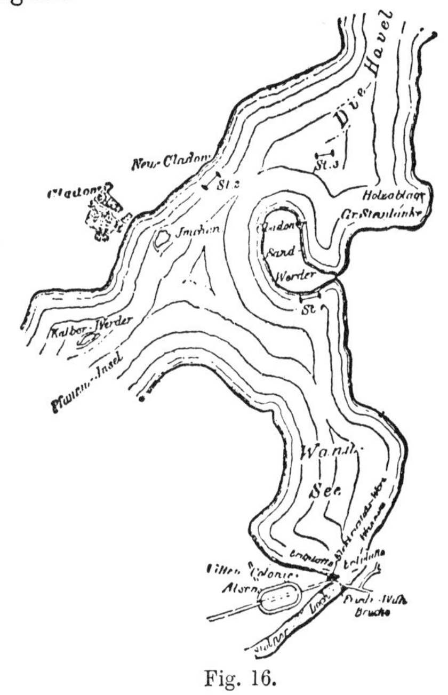

Fig. 16.

"I may point out that the resistance of the receiving circuit should be as small as possible. At first I found it difficult to produce a call in the distant receivers, but this apparently knotty problem may be solved by attaching a microphone to the membrane of the receiver, which, acting upon a relay in a local circuit, produces the call.

"It does not seem necessary to point out that by the

use of several current generators, each one producing a definite number of current impulses, a number of non-interfering messages may be sent through the water to distant telephones, each being constructed to respond to but one definite rate of vibration; or by means of one current generator a message may be sent (simultaneously) to several distant telephone receivers.

"The usefulness of this method of transmission would be much increased if means can be found to produce a written message. On the suggestion of Dr Rubens an apparatus is now being constructed, generally on the plan of Dr Wien's optical telephone. It is expected that the use of this apparatus will enable us to transform the acoustical into optical signals, and to register these photographically."

Fig. 16 shows the locality of these experiments. It will be noticed that a large sandbank intervenes between the stations, but without any appreciable effect on the results.

Prof. Rathenau concludes a very interesting paper with the enumeration of the chief points to be observed for increasing the effective signalling distance:—

"1. Great current strength in the primary circuit.

"2. Increasing the distance between the primary electrodes.

"3. Increasing the distance between the receiving electrodes.

"4. Replacing the metallic diaphragm of the telephone receiver by a light tongue.

"5. Which should be tuned to respond to a definite rate of vibration."[1]

[1] Experiments, based on the same conductive principle, were tried in Austria about the same time, but with what success I cannot say, as the results, for military reasons, have not been published.

THIRD PERIOD—THE PRACTICAL.

SYSTEMS IN ACTUAL USE.

> "The invention all admired; and each how he
> To be the inventor missed—so easy seemed
> Once found, which yet unfound most would have thought
> Impossible."

SIR W. H. PREECE'S METHOD.

SIR WM. PREECE, lately the distinguished engineer-in-chief of our postal telegraphs, has made the subject of wireless telegraphy a special study for many years, his first experiment dating back to 1882.[1] From that year up to the present he has experimented largely in all parts of the country, and has given us the results in numerous papers—so numerous, in fact, that they offer a veritable *embarras des richesses* to the historian. In what follows I can only attempt a *résumé*, and that a condensed one; but to the reader greatly interested in the subject I would advise a careful study of all the papers, a list of which I append:—

1. Recent Progress in Telephony: British Association Report, 1882.

---

[1] Indeed, it so happens that one of the first experiments he ever made in electricity was on this very subject in 1854. See p. 28, *supra*.

2. On Electric Induction between Wires and Wires: British Association Report, 1886.
3. On Induction between Wires and Wires: British Association Report, 1887.
4. On the Transmission of Electric Signals through Space: Chicago Electrical Congress, 1893.
5. Electric Signalling without Wires: Journal of the Society of Arts, February 23, 1894.
6. Signalling through Space: British Association Report, 1894.
7. Telegraphy without Wires: Toynbee Hall, December 12, 1896.
8. Signalling through Space without Wires: Royal Institution, June 4, 1897.
9. Ætheric Telegraphy: Institution of Electrical Engineers, December 22, 1898.
10. Ætheric Telegraphy: Society of Arts, May 3, 1899.[1]

In his first-quoted paper of 1882, speaking of disturbances on telephone lines, Sir William says: "The discovery of the telephone has made us acquainted with many strange phenomena. It has enabled us, amongst other things, to establish beyond a doubt the fact that electric currents actually traverse the earth's crust. The theory that the earth acts as a great reservoir for electricity may be placed in the physicist's waste-paper basket, with phlogiston, the materiality of light, and other old-time hypotheses. Telephones have been fixed upon a wire passing from the ground floor to the top of a large building (the gas-pipes being used in place of a return wire), and Morse signals, sent from a telegraph office 250 yards distant, have been distinctly read. There are several cases on record of telephone circuits miles away from any telegraph wires, but in a line with the earth terminals, picking up telegraphic signals; and when an electric-light system uses the earth, it is stoppage to all

[1] This list does not pretend to be complete. Doubtless there are other papers, which have escaped my notice.

telephonic communication in its neighbourhood. Thus, communication on the Manchester telephones was not long ago broken down from this cause; while in London the effect was at one time so strong as not only to destroy all correspondence, but to ring the telephone-call bells. A telephone system, using the earth in place of return wires, acts, in fact, as a shunt to the earth, picking up the currents that are passing in proportion to the relative resistances of the earth and the wire."[1]

He then describes the experiment which he had recently (March 1882) made of telegraphing across the Solent, from Southampton to Newport in the Isle of Wight, without connecting wires. "The Isle of Wight," he says, "is a busy and important place, and the cable across at Hurst Castle is of consequence. For some cause the cable broke down, and it became of great importance to know if by any means we could communicate across, so I thought it a timely opportunity to test the ideas that had been promulgated by Prof. Trowbridge. I put a plate of copper, about 6 feet square, in the sea at the end of the pier at Ryde (fig. 17). A wire (overhead) passed from there to Newport, and thence to the sea at Sconce Point, where I placed another copper plate. Opposite, at Hurst Castle, was a similar plate, connected with a wire which ran through Southampton to Portsmouth, and terminated in another plate in the sea at Southsea Pier. We have here a complete circuit, if we include the water, starting from Southampton to Southsea Pier, 28 miles; across the sea, 6 miles; Ryde through Newport to Sconce Point, 20 miles; across the water again, 1¼ mile; and Hurst Castle back to Southampton, 24 miles.

"We first connected Gower-Bell loud-speaking telephones in the circuit, but we found conversation was impossible. Then we tried, at Southampton and Newport, what are

[1] For early notices of the same kind, see pp. 74-80, *supra*.

called *buzzers* (Theiler's Sounders)—little instruments that make and break the current very rapidly with a buzzing sound, and for every vibration send a current into the circuit. With a *buzzer*, a Morse key, and 30 Leclanché cells at Southampton, it was quite possible to hear the

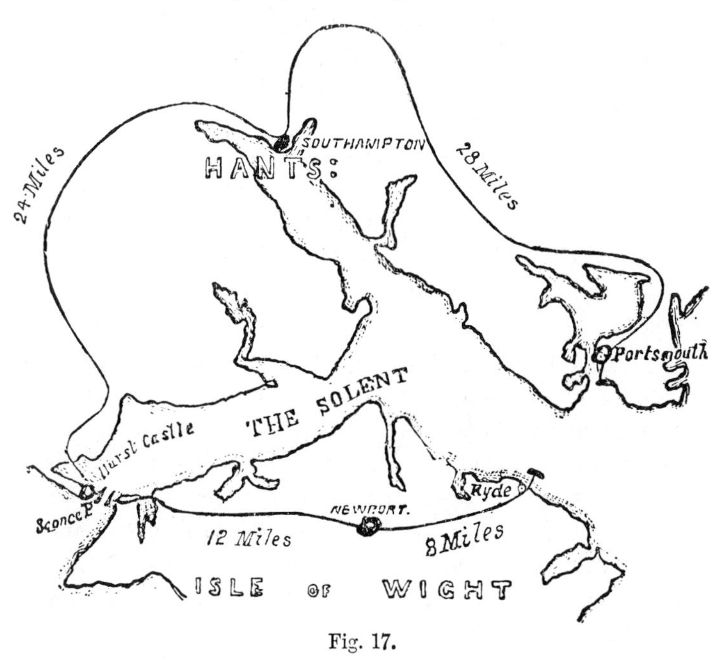

Fig. 17.

Morse signals in a telephone at Newport, and *vice versâ*. Next day the cable was repaired, so that further experiment was unnecessary."[1]

Preece, however, kept the subject in view, and in 1884 he began a systematic investigation, theoretically and experi-

[1] Captain (now Colonel) Hippisley, R.E., who conducted these trials, thought that the presence of the broken cable across the Solent somewhat vitiated the results, as its heavy iron sheathing may have aided in conducting the current.

mentally, of the laws and principles involved—an investigation which he has hardly yet completed. In his papers read at the International Electrical Congress, Chicago, August 23, 1893, and at the Society of Arts, London, February 23, 1894, he gives a *résumé* of his experiments from 1884 to date.

He begins the latter paper by asking the same momentous question which a lady once put to Faraday, What is electricity? Faraday, with true philosophic caution, replied (I quote from memory): "Had you asked me forty years ago, I think I would have answered the question; but now, the more I know about electricity, the less prepared am I to tell you what it is." Sir William is not quite so epigrammatic, nor nearly so cautious; but, then, we have learned a great deal since Faraday's time. "Few," he says, "venture to reply boldly to this question—first, because they do not know; secondly, because they do not agree with their neighbours, even if they think they know; thirdly, because their neighbours do not agree among themselves, even as to what to apply the term.[1] The physicist applies it to one thing, the engineer to another. The former regards his electricity as a form of ether, the latter as a form of energy. I cannot grasp the concept of the physicist, but electricity as a form of energy is to me a concrete fact. The electricity of the engineer is something that is generated and supplied, transformed and utilised, economised and wasted, meted out and paid for. It produces motion of matter, heat, light, chemical decomposition, and sound; while these effects are reversible, and sound, chemical decomposition, light, heat, and motion reproduce those effects which are called electricity."

[1] "Substantialists" call it a kind of matter. Others view it as a form of energy. Others, again, reject both these views. Prof. Lodge considers it a form, or rather a mode of manifestation, of the

In experiments of this kind it is necessary to point out that if we have two parallel conductors, separated from each other by a finite space, and each forming part of a separate and distinct circuit, either wholly metallic or partly completed by the earth, and called respectively the *primary* and the *secondary* circuits, we may obtain currents in the latter either by conduction or by induction; and we may classify them into those due to—

1. Earth-currents or leakages.
2. Electro-static induction currents.
3. Electro-magnetic induction currents.

It is very important to eliminate (1), which is a case of conduction, from (2) and (3), which are cases of induction, pure and simple.

1. *Earth-currents or Leakages.*

When a linear conductor dips at each end into the earth, and voltage is impressed upon it by any means, the resulting return current would probably flow through the earth in a straight line between these two points if the conductibility of the earth were perfect; but as the earth, *per se*, is a very poor conductor (and probably is so only because it is moist), lines of current-flow spread out symmetrically in a way that recalls the figure of a magnetic field. These diffused currents are evident at great distances, and can be easily traced by means of exploring earth-plates or rods. The primary current is best produced by alternating currents of such a frequency as to excite a distinct musical

ether. Prof. Nikola Tesla demurs to this view, but sees no objection to calling electricity ether associated with matter, or bound ether. High authorities cannot even yet agree whether we have one electricity or two opposite electricities.—Sir W. Crookes, 'Fortnightly Review,' February 1892.

note in a telephone, and if these currents rise and fall periodically and automatically, they produce an unmistakable wail, which, if made and broken by a Morse key into short and long periods, can be made to represent the dots and dashes of the Morse alphabet. The secondary circuit, which contains the receiving telephone, is completed in the case of an earth area by driving two rods into the ground, and in the case of water by dipping plates therein, 5 to 10 yards apart.

It is therefore necessary to be able to distinguish these earth-currents from those due to induction, as they are apt to give false effects, and to lead to erroneous conclusions. This is easily done, if the instrument be sensitive enough, by making the primary current continuous when the earth-current also becomes continuous, whereas the induction currents will be momentary, and will only be observed at the beginning and end of the primary or inducing current.

2. *Electro-static Induction Currents.*

When a body, A, is electrified by any means and isolated in a dielectric, as air, it establishes an electric field about it; and if in this field a similar body, B, be placed, it also is electrified by induction. If B be placed in connection with the earth, or with a condenser, or with any very large body, a charge of the same sign as A is conveyed away, and it (B) remains electrified in the opposite sense to A. A and B are now seats of electric force or stress. The dielectric between them is displaced or, as we say, *polarised*—that is, it is in a state of electric strain, and remains so as long as A remains charged; but if A be discharged, or have its charge reversed or varied, then similar changes occur in B, and in the dielectric separating them. A may be an extended wire forming part of a complete primary circuit, and its charge may be due to a battery or other source of electricity; then, in the equally extended secondary wire B (fig. 18), the displaced charge in flowing to earth establishes a momentary current whose direction and duration depend on the current in A, and on its rate of variation.

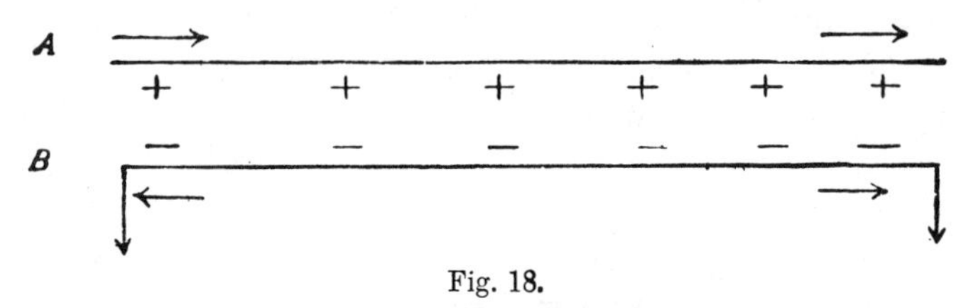

Fig. 18.

The strained (polarised) state of the dielectric, and the charges on A and B, remain quiescent so long as the current flows steadily; but when it ceases we have again, and in *both* circuits, momentary currents, as shown by the arrows (fig. 19), which flow until equilibrium is restored.

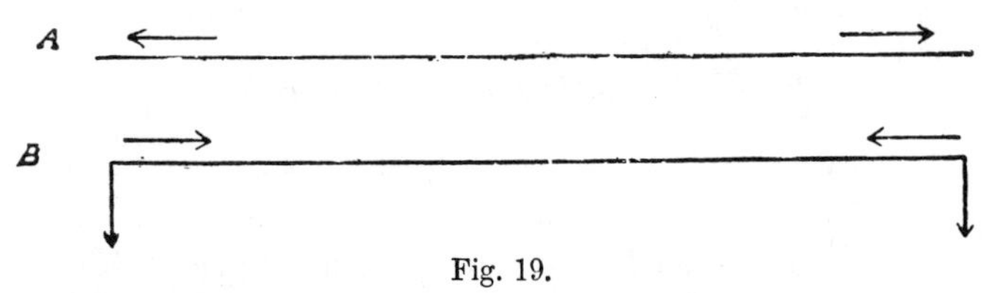

Fig. 19.

The secondary currents due to discharge, like those due to charge, flow in opposite directions at each end, and there is always some intermediate zero point.

It is thus easy in long circuits, by observing their direction, to differentiate currents of induction due to electrostatic displacement from those due to electro-magnetic disturbance.

The effects of electro-static induction do not play an important part in the inquiry immediately before us, but they are of great consequence in questions of speed of signalling in submarine cables and long, well-insulated land-lines, and in clearness of speech in long-distance telephony.[1]

3. *Electro-magnetic Induction Currents.*

Magnetic force is that which produces, or tends to produce, polarisation in magnetisable matter (as iron, nickel, cobalt), and electro-magnetic disturbance or stress in non-magnetisable matter and the ether. An electric current in a conductor is a seat of magnetic force, and establishes in its neighbourhood a magnetic field. The lines of force in this field are equivalent to circles in a plane perpendicular to the direction of the current, which circles, during the rise of the current, flow outwards or expand, and, during the fall of the current, flow inwards or contract, much like the waves on the surface of smooth water when a pebble is thrown in, but moving with the speed of light Thus any linear conductor placed in the field of another parallel conductor carrying a current is cut at right angles to itself by these lines of force—in one direction as the current rises, and in the opposite direction as the current falls. This outward and inward projection of magnetic force through such linear conductor excites electric force in that conductor, and if it form part of a circuit an electric current is set up in that circuit.

So far for the theory of the subject. Now for its experimental elucidation. Besides those cases of interference mentioned on p. 136, others were of frequent occurrence in the experience of the postal-telegraph officials, the most striking being that known as the Gray's Inn Road case. In 1884 it was there noticed that messages sent in the ordinary

[1] For an interesting investigation of electro-static phenomena on telephone circuits, see Mr Carty's papers in the 'Electrician,' December 6, 1889, and April 10, 1891.

way through insulated wires, buried in iron pipes along the road, could be read upon telephone circuits erected on poles on the house-tops 80 feet high. To cure the evil the telegraph wires had to be taken up and removed to a more distant route.[1]

In 1885 Preece arranged an exhaustive series of experiments in the neighbourhood of Newcastle, which were ably carried out by Mr A. W. Heaviside, to determine whether these disturbances were due to electro-magnetic induction, and were independent of earth conduction; and also to find out how far the distance between the wires could be extended before this influence ceased to be evident. Insulated squares of wire, each side being 440 yards long, were laid out horizontally on the ground one quarter of a mile apart, and distinct speech by telephones was carried on between them; while when removed 1000 yards apart inductive effects were still appreciable.

With the parallel lines of telegraph, ten and a quarter miles apart, between Durham and Darlington, the ordinary working currents in the one were clearly perceptible in a telephone on the other. Even indications were obtained in this way between Newcastle and Gretna, on the east and west coasts, forty miles apart; but here the observations were doubtless vitiated by conduction or leakage through

[1] The following are more recent cases of the same kind. Currents working the City and South London Electric Railway affect recording galvanometers at the Greenwich Observatory, four and a half miles distant; and even a diagram of the train service could be made out by tapping any part of the metropolitan area.

Some ten years ago one of the dynamos at the Ferranti electric-light station at Deptford by some accident got connected to earth, with the result that the whole of the railway telegraphs in the signal-boxes of the railways in South London were temporarily put out of order and rendered inoperative, while the currents flowing in the earth were perceived in the telegraph instruments so far northwards as Leicester and so far south as Paris.

the large network of telegraph wires between those two places.[1]

The district between Gloucester and Bristol, along the banks of the Severn, was next (1886) selected, where for a length of fourteen miles, and an average distance apart of four and a half miles, no intermediate disturbing lines existed. Complete metallic circuits were employed, the return wires passing far inland, in the one case through Monmouth, and in the other through Stroud. In one wire currents of about ·5 ampère were rapidly made and broken by mechanical means, producing on a telephone a continuous note which could be broken up by a Morse key into dots and dashes, as in Cardew's vibrator. Weak disturbances were detected in the secondary circuit, showing that here the range of audibility with the apparatus in use was just overstepped. The unexpected fact was also shown in these experiments that, whether the circuits were entirely metallic or earthed at the ends, the results were the same.[2]

Similar trials were made on lines along the valley of the Mersey. A new trunk line of copper wires that was being erected between London and the coast of North Wales was then experimented upon, and some interesting results were obtained in the district between Shrewsbury and Much Wenlock, and between Worcester and Bewdley.

In the autumn of the same year (1886) some admirable results were obtained by Mr Gavey, another of Preece's able assistants, near Porthcawl, in South Wales—a wide expanse of sand well covered by the tide, thus giving the opportunity of observing the effects in water as well as in air. Two horizontal squares of insulated wire, 300 yards each side, were laid side by side at various distances apart

[1] British Association Report, 1886.

[2] These experiments were repeated with more experience and greater success in 1889.

up to 300 yards, and the inductive effects of one on the other noted. Then one coil was suspended on poles 15 feet above the other, which was covered with water at high tide. No difference was observable in the strength of the induced signals, whether the intervening space was air or water or a combination of both, although subsequent experience (1893) showed that with a space of 15 feet the effect in air was distinctly better than through water.

The conclusion drawn from all these experiments was that the magnetic field extends uninterruptedly through the earth, as it does through the air; and that if the secondary circuit had been in a coal-pit the effect would be the same. In fact, Mr Arthur Heaviside succeeded in 1887 in communicating between the surface and the galleries of Broomhill Colliery, 350 feet deep. He arranged a circuit in a triangular form along the galleries about two and a quarter miles in total length, and at the surface a similar circuit of equal size over and parallel to the underground line. Telephonic speech was easily carried on by induction from circuit to circuit.[1]

As the result of all these experiments and innumerable laboratory investigations, Preece deduced the following formulæ. The first shows the strength of current C_2 induced in the secondary circuit by a given current C_1 in the primary circuit—

$$C_2 = \frac{C_1}{R} \frac{\sqrt{L^2 + D^2} - D}{D} \times I,$$ [2]

[1] Subsequent experiments showed that the conclusion arrived at for earth and air was only partially true for water. Telephonic speech was carried on in Dover Harbour through 36 feet of water, but no practical signals could be obtained through 400 feet at North Sand Head, Goodwin Sands, showing that the effect must diminish in water with some high power of the distance.

[2] This formula does not allow for the reverse effect of the return current through the earth, as to which no data exist at present.

where R equals the resistance of the secondary circuit, D the distance apart of the two circuits, L the length of the inductive system, and I the inductance of the system. The value of I, obtained by experiment on two parallel squares of wire, 1200 yards round and 5 yards apart, was found to be ·003.

The second equation gives approximately the maximum distance X which should separate any two wires of length L, C_1 being the primary current and R the resistance of the secondary circuit.

$$X = 1{\cdot}9016 \frac{\sqrt{C_1 L}}{R}$$

The constant 1·9016 was obtained by experimenting on two parallel wires, each one mile long, when the primary circuit, being excited by one ampère, the limit of audibility in the secondary was reached at 1·9016 miles. This formula shows the desirability of using copper wires of the largest size practicable, so as to reduce the value of R. Other very important elements of success are (1) the rate at which the primary currents rise and fall, the faster the better, and (2) the reduction to a minimum of such retarding causes as capacity and self-induction.

Having thus threshed out the laws and conditions of electro-magnetic disturbances, and determined the distance at which they could be usefully applied, it only remained for Sir William to put his conclusions to a practical test. Accordingly, when the Royal Commission on electric communication between the shore and lighthouses and lightships was appointed in June 1892, he made his proposals to the Government, and on receiving sanction forthwith proceeded to carry them out.

The Bristol Channel proved a very convenient locality to test the practicability of communicating across distances of three and five miles without any intermediate conductors. Two islands, the Flat Holm and the Steep Holm, lie off Penarth and Lavernock Point, near Cardiff, the former having a lighthouse upon it (fig. 20). On the shore two thick copper wires combined in one circuit were suspended on poles for a distance of 1267 yards, the circuit being

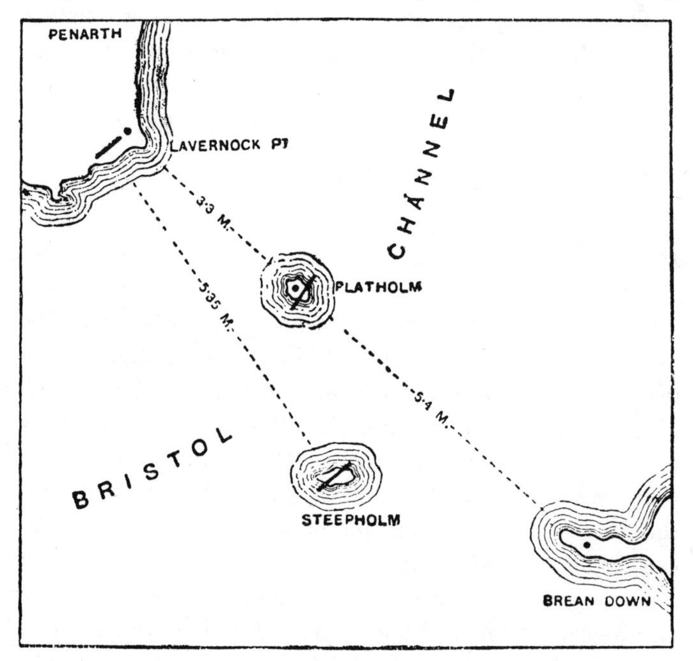

Fig. 20.

completed by the earth. On the sands at low-water mark, 600 yards from this primary circuit and parallel to it, two gutta-percha covered copper wires and one bare copper wire were laid down, their ends being buried in the ground by means of bars driven in the sand.

One of the gutta-percha wires was lashed to an iron wire to represent a cable. These wires were periodically covered by the tide, which rises here at spring to 33 feet. On the Flat Holm, 3·3 miles away, another gutta-percha covered copper wire was laid for a length of 600 yards.

There was also a small steam launch having on board several lengths of gutta-percha covered wire. One end of such a wire, half a mile long, was attached to a small buoy, which acted as a kind of float to the end, keeping the wire suspended near the surface of the water as it was paid out while the launch slowly steamed ahead against the tide. Such a wire was paid out and picked up in several positions between the primary circuit and the islands.

The apparatus used on shore was a 2-h.p. portable Marshall's engine, working a Pyke and Harris's alternator, sending 192 complete alternations per second of any desirable strength up to a maximum of 15 ampères. These alternating currents were broken up into Morse signals by a suitable key. The signals received on the secondary circuits were read on a pair of telephones—the same instruments being used for all the experiments.

The object of the experiments was not only to test the practicability of signalling between the shore and the lighthouse, but to differentiate the effects due to earth conduction from those due to electro-magnetic induction, and to determine the effects in water. It was possible to trace without any difficulty the region where they ceased to be perceptible as earth-currents and where they commenced to be solely due to electro-magnetic waves. This was found by allowing the paid-out cable, suspended near the surface of the water, to sink. Near the shore no difference was perceptible, whether the cable was near the surface or lying on the bottom, but a point was reached, just over a mile away, where all sounds ceased as the cable sank, but were received again when the cable came to the surface. The total absence of sound in the submerged cable was rather surprising, and led to the conclusion either that the electro-magnetic waves of energy are dissipated in the sea-water, which is a conductor, or else that they are reflected away from the surface of the water, like rays of light.[1]

Experiments on the Conway Estuary, showing the relative transparency of air and water to these electro-magnetic waves, tend to support the latter deduction; for if much waste of energy took place in the water, the difference would be more marked. As it is, there seems to be ample evidence that the electro-magnetic waves are transmitted to considerable distances through water, though how far remains to be found.[2]

There was no difficulty in communicating between the shore and Flat Holm, 3·3 miles. The attempt to speak between Lavernock and Steep Holm, 5·35 miles, was not so successful: though signals were perceptible, conversation was impossible. There was distinct evidence of sound, but it was impossible to differentiate the sounds into Morse signals. If either line had been longer, or the primary currents stronger, signalling would probably have been possible.

In 1894 Preece carried out some satisfactory experiments near Frodsham, on the estuary of the Dee, which was found to be a more convenient locality than the Conway sands. Here, as at Conway and other places, squares and rectangles were formed of insulated wires, and numerous measurements were made (with reflecting galvanometers and telephones) of the effects due to varying currents in the primaries, and at varying distances between them and the secondaries.

In Scotland also some very successful trials were made. There happens to be a very convenient and accessible loch

[1] See note, p. 167, *infra*. [2] See note, p. 146, *supra*.

in the Highlands—Loch Ness—forming part of the route of the Caledonian Canal between Inverness and Banavie, having a line of telegraph on each side of it. Five miles on each side of this loch were taken, and so arranged that any fractional length of telegraph wire on either side could be taken for trial. Ordinary, and not special, apparatus was employed. Sending messages, as before, by Morse signals and speaking by telephone across a space of one and a quarter mile was found practical, and, in fact, easy; indeed, the sounds were so loud that they were found sufficient to form a call for attention.

The following apparatus was in use on each side of the loch: A set of batteries consisting of 100 dry cells, giving a maximum voltage of 140; a rapidly revolving rheotome, which broke up the current into a musical note; a Morse key, by which these musical notes could be transformed into Morse signals; resistance coils and ampère-meters to vary the primary current; two Bell telephones joined in multiple arc to act as receivers. For the transmission of actual speech simple granular carbon microphones, known as Deckert's, were used as transmitters, and a current of two ampères was maintained through these and two Bell telephones in circuit with the line wire.

Any lingering fear that earth conduction had principally to do with these results was removed by making the earth's terminals on the primary circuit at one end at Inverness nine miles away, and at the other end in two directions in a parallel glen about six miles away.

One very interesting fact observed at Loch Ness was that there was one particular frequency in the primary circuit that gave a decided maximum effect upon the telephones in the secondary circuit. This confirms the presence of resonance, and is, of itself, a fact sufficient to prove the effects as being due to the transformation of electromagnetic waves into electric currents.[1]

During the same year (1894) experiments were carried out between the island of Arran and Kintyre across Kil-

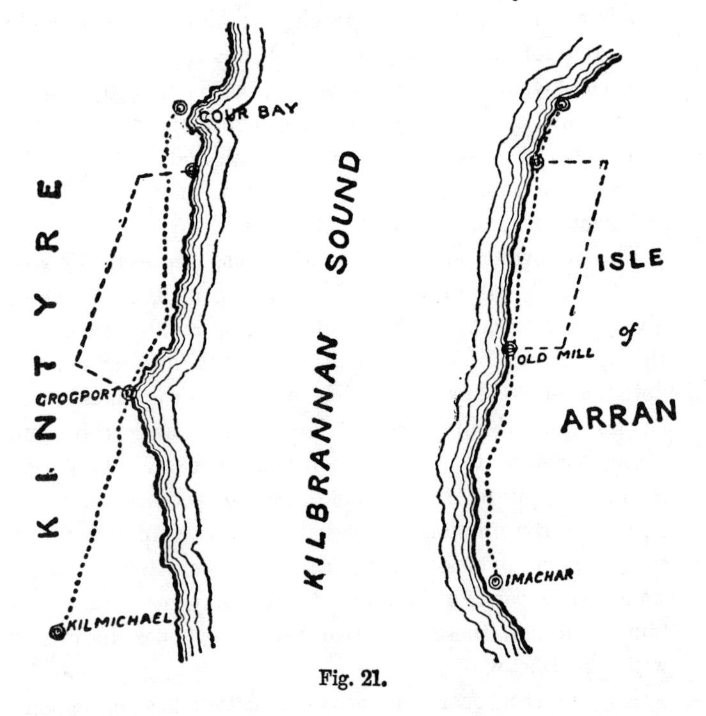

Fig. 21.

brannan Sound. Two parallel lines on opposite sides, and four miles apart, were taken (fig. 21); and, in addition, two gutta-percha covered wires were laid along each coast, at a height of 500 feet above sea-level and five miles apart horizontally.

[1] This is still a moot question, many competent authorities, as Lodge, Rathenau, W. S. Smith, and Stevenson, being of opinion that the effect is partly inductive and partly conductive. See Dr Lodge's contention, 'Jour. Inst. Elec. Engs.,' No. 137, p. 814.

Incidentally some extremely interesting effects of electro-magnetic resonance were observed during the experiments in Arran. A metallic circuit was formed partly of the insulated wire 500 feet above the sea-level and partly of an ordinary line wire, the rectangle being two miles long and 500 feet high. Wires on neighbouring poles, at right angles to the shorter side of the rectangle, *although disconnected at both ends*, took up the vibrations, and it was possible to read all that was signalled on a telephone placed midway in the disconnected circuit by the surgings thus set up.

The general conclusions arrived at as the result of these numerous and long-continued experiments may be briefly summed up as follows:[1]—

The earth acts simply as a conductor, and *per se* it is a very poor conductor, deriving its conducting property principally, and often solely, from the moisture it contains. On the other hand, the resistance of the "earth" between the two earth plates of a good circuit is practically nothing. Hence it follows that the mass of earth which forms the return portion of a circuit must be very great, for we know by Ohm's law that the resistance of a circuit increases with its specific resistance and length, and diminishes with its sectional area. Now, if the material forming the "earth" portion of the circuit were, like the sea, homogeneous, the current-flow between the earth plates would follow innumerable but definite stream lines, which, if traced and plotted out, would form a hemispheroid. These lines of current have been traced and measured. A horizontal plan on the surface of the earth is of the form illustrated in fig. 22, while a vertical section through the earth is of the form shown in fig. 23.

With earth plates 1200 yards apart these currents have

[1] British Association Report, 1894, Section G.

been found on the surface at a distance of half a mile behind each plate; and, in a line joining the two trans-

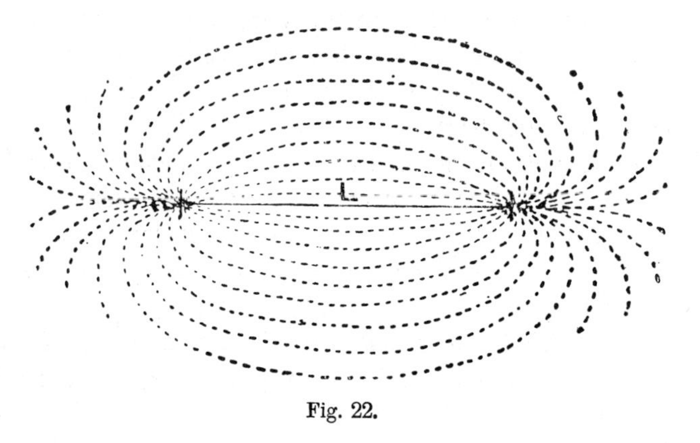

Fig. 22.

versely, they are evident at a similar distance at right angles to this line.

Now this hemispheroidal mass could be replaced electrically by a resultant conductor (R, fig. 23) of a definite form

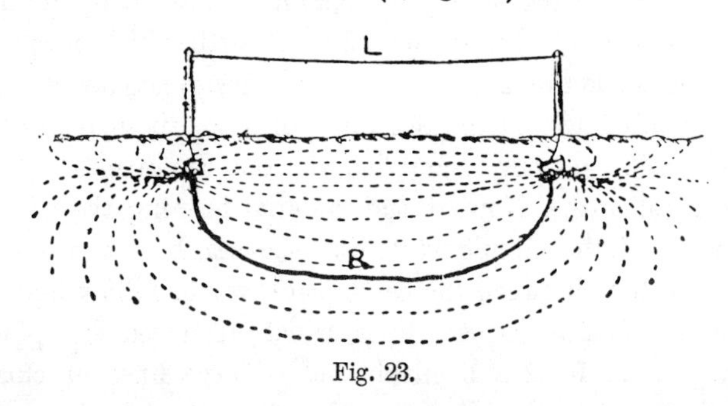

Fig. 23.

and position; and, in considering the inductive action between two circuits having earth returns, it is necessary to estimate the position of this imaginary conductor. This was the object of the experiments at Frodsham.

If the material of the earth be variable and dry the hemispheroid must become very much deformed and the section very irregular: the lines of current-flow must spread out farther, but the principle is the same, and there must be a resultant return. The general result of the experiments at Frodsham indicates that the depth of the resultant earth was 300 feet, while those at Conway are comparable with a depth of 350 feet. In the case of Frodsham the primary coil had a length of 300 feet, while at Conway the length was 1320 feet. At Loch Ness, and between Arran and Kintyre, where the parallel lines varied from two to four miles, the calculated depth was found to be about 900 feet. The depth of this resultant must, therefore, increase with the distance separating the earth plates, and this renders it possible to communicate by induction from parallel wires over much longer distances than would otherwise be possible.

The first and obvious mode of communicating across space is by means of coils of wire opposed to each other in the way familiar to us through the researches of Henry and Faraday. All the methods here described consisted in opposing two similar coils of wire having many turns, the one coil forming the primary circuit and the other coil the secondary circuit.

Vibratory or alternating currents of considerable frequency were sent through the primary circuit, and the induced secondary currents were detected by the sound or note they made on a telephone fixed in the secondary circuit.

The distance to which the effective field formed by a coil extends increases with the diameter of the coil more than with the number of turns of wire upon it. A single wire stretched across the surface of the earth, forming part of a circuit completed by the earth, is a single coil, of which the lower part is formed by the resultant earth return, and the distance to which its influence extends depends upon the height of the wire above the ground and the depth of this resultant earth.

In establishing communication by means of induction, there are three dispositions of circuit available—viz., (*a*) single parallel wires to earth at each extremity; (*b*) parallel coils of one or more turns; (*c*) coils of one or more turns placed horizontally and in the same plane.

The best practical results are obtained with the first arrangement, more especially if the conformation of the earth admits of the wires being carried to a considerable height above the sea, whilst the earth plates are at the sea-level. By adopting this course the size of the coil is practically enlarged, and even if it be necessary to increase the distance between the parallel wires in order to get a larger coil, the result is still more beneficial. In a single-wire circuit we have the full effect of electro-static and electro-magnetic induction, as well as the benefit of any earth conduction, but in closed coils we have only the electro-magnetic effects to utilise.

In one experiment two wires of a definite length were first made up into two coils forming metallic circuits, then uncoiled and joined up as straight lines opposed to each other, with the circuit completed by earth. The effects, and the distance between which they were observable, were very many times greater with the latter than with the former arrangement.

The general law regulating the distance to which we can speak by induction has not been *rigorously* determined, and it is hardly possible that it can be done, owing to the many disturbing elements, geological as well as electrical. In practice we have to deal with two complete circuits of unknown shape, and in different planes. We have obtained some remarkably concordant and accurate results in one place; but, on the other hand, we have met with equally discordant results in another place. Still, from the approximate formula before given, we deduce the important fact that for parallel lines the limiting distance increases directly as the square of the length, which shows that if we can make the length of the two lines long enough it would be easy to communicate across a river or a channel. Of course, as previously pointed out, the formula does not take into account the effects of the reverse magnetic waves generated by the return current through the earth, and at present no data exist on which a satisfactory calculation can be based; but, for example, there is little doubt that two wires, ten miles long, would signal through a distance of ten miles with ease.

"Although," says Sir William in conclusion, "communication across space has thus been proved to be practical in certain conditions, those conditions do not exist in the cases of isolated lighthouses and light-ships, cases which it was specially desired to provide for. The length of the secondary must be considerable, and, for good effects, at least equal to the distance separating the two conductors. Moreover, the apparatus to be used on each circuit is cumbrous and costly, and it may be more economical to lay an ordinary submarine cable.

"Still, communication is possible even between England and France, across the Channel, and it may happen that between islands where the channels are rough and rugged, the bottom rocky, and the tides fierce, the system may be financially possible. It is, however, in time of war that it may become useful. It is possible to communicate with a beleaguered city either from the sea or on the land, or between armies separated by rivers, or even by enemies.

"As these waves are transmitted by the ether, they are independent of day or night, of fog, or snow, or rain, and therefore, if by any means a lighthouse can flash its indicating signals by electro-magnetic disturbances through space, ships could find out their positions in spite of darkness and of weather. Fog would lose one of its terrors, and electricity become a great life-saving agency."

At the Society of Arts (February 23, 1894), Sir William gave rein to his imagination, and, looking beyond these mundane utilities, concluded his address with the following magnificent peroration :—

"Although this short paper is confined to a description of a simple practical system of communicating across terrestrial space, one cannot help speculating as to what may occur through planetary space. Strange mysterious sounds are heard on all long telephone lines when the earth is used as a return, especially in the calm stillness of night. Earth-currents are found in telegraph circuits and the aurora borealis lights up our northern sky when the sun's photosphere is disturbed by spots. The sun's surface must at such times be violently disturbed by electrical storms, and if oscillations are set up and radiated through space, in sympathy with those required to affect telephones, it is not a wild dream to say that we may hear on this earth a thunderstorm in the sun.

"If any of the planets be populated with beings like ourselves, having the gift of language and the knowledge to adapt the great forces of nature to their wants, then, if they could oscillate immense stores of electrical energy to and fro in telegraphic order, it would be possible for us to hold commune by telephone with the people of Mars."

The first application of Preece's system to the ordinary needs of the postal-telegraph service was made on March 30, 1895, when the cable between the Isle of Mull and Oban, in Scotland, broke down. As there was no ship available at the moment for effecting repairs, communication was established by laying a gutta-percha-covered copper wire, one and a half mile long, along the ground from Morven, on the Argyllshire coast, while on Mull the ordinary telegraph (iron) wire connecting Craignure with Aros was used, the mean distance separating the two base lines being about two miles. No difficulty was experienced in keeping up communication, and many public messages were transmitted for a week until the cable was repaired. In all about 160 messages were thus exchanged, including a press telegram of 120 words.

The diagram (fig. 24) shows the apparatus and connec-

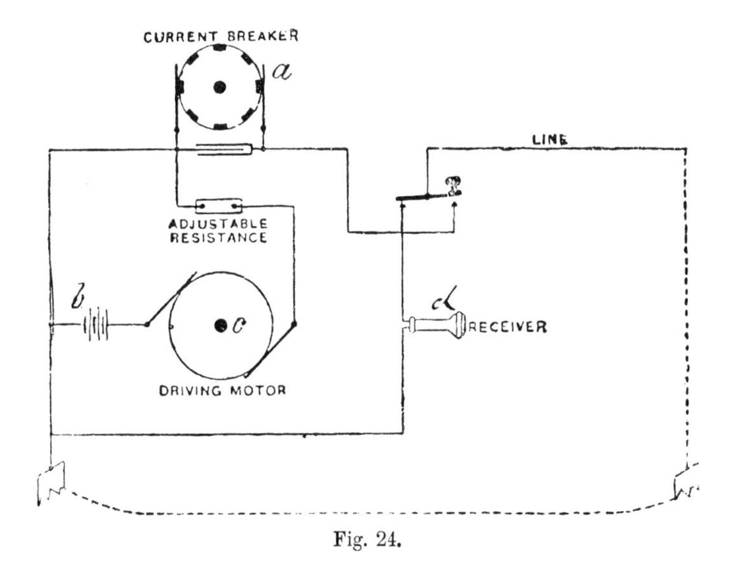

Fig. 24.

tions, as regards which it is only necessary to say that *a* is a rheotome, or make-and-break wheel, driven so as to produce about 260 interruptions of the current per second, which give a pleasant note in the telephone, and are easily read when broken up by the key into Morse dots and dashes; *b* is a battery of 100 Leclanché cells, of the so-called dry and portable type; *c* is a switch to start and stop the rheotome as required; and *d* is a telephone to act as receiver.

Since March 1898 this system has been permanently established for signalling between Lavernock Point and the Flat Holm, and has been handed over to the War Office. Permanent lines of heavy copper wire have been erected parallel to each other, one being on the Flat Holm and the other on the mainland.

The heavy and cumbrous Pyke and Harris alternator of the earlier experiment over the same line (p. 149, *ante*) has been replaced by 50 Leclanché cells. The frequency has been raised to 400 makes and breaks per second, thus greatly increasing the strength of the induced currents. By the use of heavy copper base lines the resistances have been made as low as practicable. There is no measurable capacity, self-induction is eliminated, and there is no impedance. Hence the signals are perfect, and the rate of working is only dependent on the skill of the operator. It is said that as many as 40 words per minute have been transmitted without the necessity for a single repetition—a speed which few telegraphists can achieve, and still fewer can keep up.

A little later Mr Sydney Evershed's relays were added to work a call-bell, which was the only thing wanted to make the system complete and *practical*.[1]

[1] During the summer of 1899 Sir William began a new series of experiments on wireless telephony at the Menai Straits, the results of which he communicated recently to the British Association (Bradford, September 8, 1900). After referring to his Loch Ness experiments (p. 151, *ante*), where telephonic signals were found possible across an average space of 1·3 miles with parallel base lines of 4 miles each, Sir William states that his new experiments fully bore out this fact, and determined the further fact that maximum effects are obtained when the parallel wires are terminated by earth-plates in the sea itself—showing that the inductive effects through the air are enhanced by conductive effects through the water, and that, consequently, shorter base lines are permissible. Ordinary telephone transmitters and receivers were used.

This new method has been successfully applied to establishing communication between the Skerries and Cemlyn, Anglesey, across 2·8 miles average distance, and between Rathlin Island and the Irish coast, about 4 miles across.

WILLOUGHBY SMITH'S METHOD.

Mr Smith's researches in wireless telegraphy date back to 1883. His first suggestions, of the induction order, were contained in a paper on Voltaic-Electric Induction, which he read before the Institution of Electrical Engineers on November 8 of that year. These have already been noticed in our account of Edison's invention (p. 102, *supra*).

Somewhat later, early in 1885, Mr Smith turned his attention to conduction methods, and worked out a plan which, in a modified form, has been in actual operation for the last three years.

The *rationale* of the system is described by Mr Smith as follows:—

"Messages have been sent and correctly received through a submarine cable two thousand miles in length, the earth being the return half of the circuit, by the aid of the electricity generated by means of an ordinary gun-cap containing one drop of water; and small though the current emanating from such a source naturally was, yet I believe it not only polarised the molecules of the copper conductor, but also in the same manner affected the whole earth through which it dispersed on its way from the outside of the gun-cap to its return, through the cable, to the water it contained. I further believe that the time will come, perhaps sooner than may be expected, when it will be possible to detect even such small currents in any part of the world in the same way that it is now possible to do in comparatively small sections of it.

"For researches of this description it is necessary to employ as sensitive an instrument as it is possible to obtain, to pick up, so to speak, such minute currents. Now, there is that wonderful instrument the telephone. I say wonderful advisedly, for as far as I know it is not to be equalled for the simplicity of its mechanical construction and the ease with which it can be manipulated, and yet is so peculiarly sensitive. I have used it in most of my experiments as the receiving instrument, although of course there are other well-known instruments that could be employed, as all depends upon the potential of the current to be detected. The sending arrangement was either an ordinary Morse key so manipulated for a short or long time as to give the necessary sounds in the telephone to represent dots and dashes, or a double key and two pieces of mechanism giving dissimilar sounds were employed with good results. I gave much time and thought to the subject, the results of each experiment giving me much encouragement to proceed.

"Of the many experiments made I select the following, as I think it will clearly illustrate my system for telegraphing to a distant point not in metallic connection with the sending station. A wooden bathing-hut on a sandy beach made a good shore station, from which were laid two insulated copper wires 115 fathoms in length. The ends of the wires, scraped clean, were twisted round anchors, their position being marked by buoys about 100 fathoms apart, and in about 6 fathoms of water. Midway between the two a boat was anchored with a copper plate hanging fore and aft about 10 fathoms apart, and consequently about 45 fathoms from either end of the anchored shore wires. This boat represented the sea station, and, owing to the state of the sea, a very wet and lively one it proved; therefore, taking this fact into consideration, together with the crude nature of the experiment, it was remarkable with what distinctness and ease messages were passed. The last message sent from shore was, 'Thanks: that will do; pick up anchors and return.' To this the reply came from the boat, 'Understand,' and they then proceeded to carry out instructions. The boat employed was a wooden one, but it would have been much better for my purpose had it been of metal, for then I should have used it instead of one of the collecting plates, as the larger the surface of these plates the better the results obtained."[1]

This method was secured by patent, June 7, 1887, from the specification of which (No. 8159) I take the following particulars: At the present time wherever electric telegraph communication is established between the shore and a lighthouse, either floating or on a rock, at a distance from the shore, it is effected through an insulated conductor or cable. Much difficulty is, however, experienced owing to the rapid wearing of the cable, so that it is liable to break whenever a storm comes on, and when, consequently, it is most required to be in working order. By this invention communication can be effected between the sending station and the distant point without the necessity of metallic connection between them.

A in the drawing (fig. 25) is a two-conductor cable led from a signal-station B on shore towards the rock C. At a distance from the rock one of the conductors is led to a metallic plate D submerged on one side of the rock, and at such a distance from it as to be in water deep enough for it not to be affected by waves. The other conductor is led to another metallic plate E similarly submerged at a distance from the opposite side of the rock. F F are two submerged metallic plates, each opposite to the plates D and E respectively. G G are insulated conductors leading from the plates F F to a telephone of low resistance in the lighthouse H.

To communicate from the shore, an interrupter or reverser I and battery K are connected to the shore ends of the two-wire cable. The telephone in the lighthouse circuit then responds to the rapid makes and breaks or reversals of the current, so that signalling can readily be

[1] 'Electrician,' November 2, 1888.

carried on by the Morse alphabet. If a vibrating interrupter or reverser be used, a short or long sound in the telephone can be obtained by a contact key held down for short or long intervals.

A more convenient way is to use two finger-keys, one of which by a series of teeth on its stem produces a few breaks or reversals of the current, whilst the other key when depressed produces a greater number of breaks or reversals.

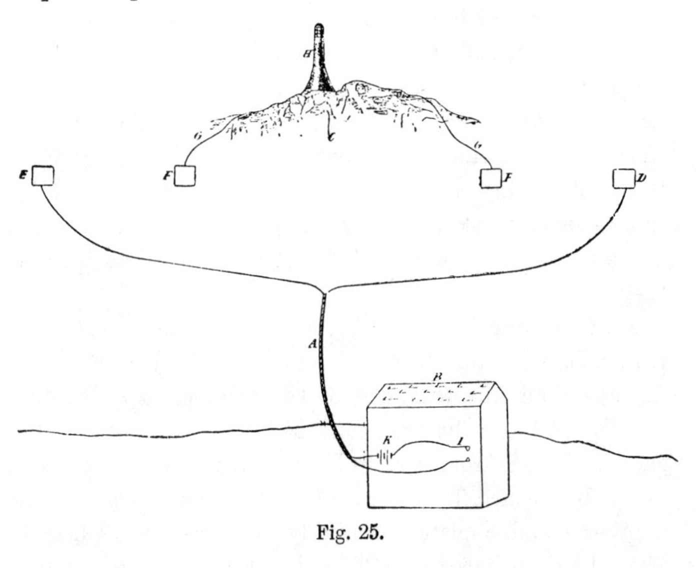

Fig. 25.

For communicating from the lighthouse to the shore a battery and make-and-break apparatus are coupled to the insulated conductors on the rock, and a telephone to the shore ends.

In the same way communication could be carried on from the shore to a vessel at a distance from it, if the vessel were in the vicinity of two submerged plates or anchors, each having an insulated conductor passing from it to the shore, and if two metallic plates were let go from the vessel so that these plates might be at a distance apart from one another. The position of the two submerged plates might be indicated by buoys. In this way communication might be effected between passing vessels and the shore, or between the shore and a moored lightship or signal-station.

A similar result might be obtained with a single insulated conductor from the shore by the use of an induction apparatus, the ends of the secondary coil being connected by insulated conductors to the submerged plates.

An important modification of this method was subsequently effected by Messrs Willoughby S. Smith & W. P. Granville,[1] based on the following reasoning:—

In fig. 26 A B represents an insulated conductor of any desired length, with ends to earth E E as shown. C is a

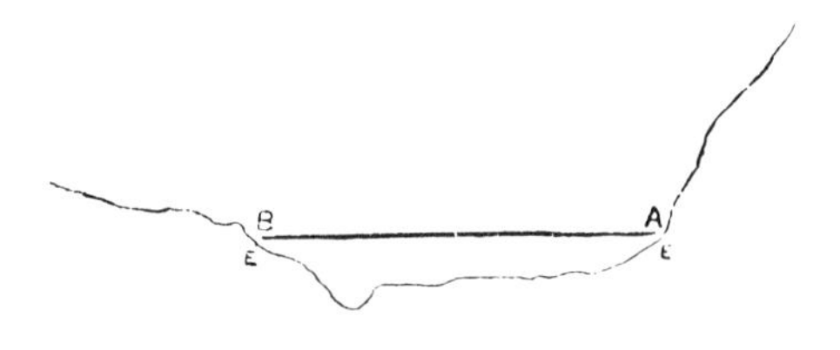

Fig. 26.

rock island on which is extended another insulated wire C D, with its ends also connected to earth. Now, if a current is caused to flow in A B, indications of it will be shown on a galvanometer in the circuit C D. This is Preece's arrangement at Lavernock and Flat Holm.

[1] See their patent specification, No. 10,706, of June 4, 1892.

Now, if we rotate the line A B round A until it assumes the position indicated in fig. 27, we have Messrs Smith & Granville's arrangement, where, owing to the proximity

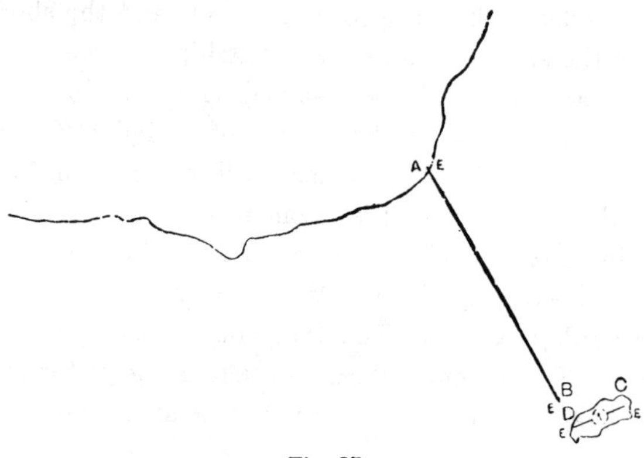

Fig. 27.

of B to D, signalling is practicable with a small battery power. Thus, where the distance from B to D was 60 yards, one Leclanché cell was found to be ample. As

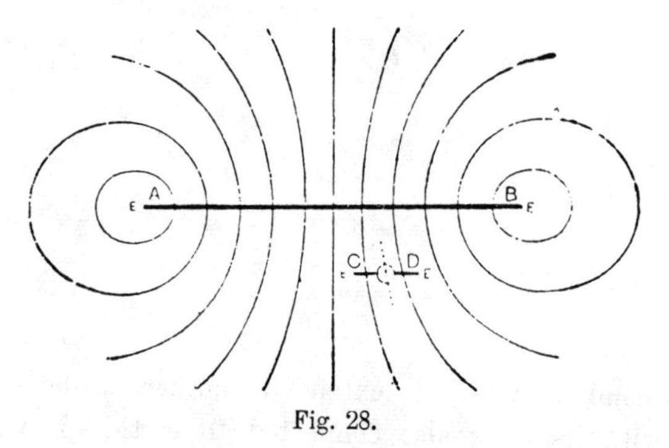

Fig. 28.

a permanent current in A B causes a *permanent* deflection on the galvanometer in C D, this deflection cannot be produced otherwise than by conduction.

Again, let A B (fig. 28) represent an insulated conductor

having its ends submerged in water (the distance between A and B being immaterial). Now cause a current to flow continuously, and it will be found that the water at each end of the conductor is charged either positively or negatively (according to the direction of the current) in equipotential spheroids, diminishing in intensity as the distance from either A or B is increased. To prove this, provide a second circuit, connected with a galvanometer, and with its two ends dipping into the water. Now, it will be found that a current flows in the C D circuit as long as the current in A B is flowing; the current in C D diminishes as C and D are moved farther away from B, and also diminishes to zero if the points C D are turned until they both lie in the same equipotential curve as shown by the dotted line.

It must be well understood that although, for the sake of clearness, the equipotential curves are shown as planes, yet in a body of water they are more or less spheres extending symmetrically around the submerged ends of the conductor, and therefore it is evident from the foregoing that the position of C D, in relation to B, must be considered not only horizontally but vertically.[1]

Early in 1892 the Trinity Board placed the Needles Lighthouse at the disposal of the Telegraph Construction and Maintenance Company, so that they might prove the practicability of the method here described. The Needles

[1] This fact, Mr Smith thinks, fully explains Preece's launch experiments (p. 149, *supra*). For instance, when the launch towing the half-mile of cable parallel to the wire on the mainland was close to the shore, the cable, although allowed to sink, could only do so to a very limited extent, and therefore still remained in a favourable position for picking up the earth-currents from A B (fig. 28); but when one mile from the shore, and in deep water, the cable was able to assume somewhat of a vertical position with the two ends brought more or less into the same equipotential sphere, it naturally resulted in a diminution or cessation of the current in the C D or launch circuit, and hence the absence of signals.

Lighthouse was chosen on account of its easy access from London.

In May 1892 an ordinary submarine cable was laid from Alum Bay to within 60 yards of the lighthouse rock, where it terminated, with its conductor attached to a specially constructed copper mushroom anchor. An earth plate close to the pier allowed a circuit to be formed through the water. On the rock itself two strong copper conductors were placed, one on either side, so that they remained immersed in the sea at low water, thus allowing another circuit to be formed through the water in the vicinity of the rock.

The telephone was first tried as the receiving instrument, with a rapid vibrator and Morse key in the sending circuit. This arrangement was afterwards abandoned, as it was not nearly so satisfactory as a mirror-speaking galvanometer, and the men, being accustomed to flag work, preferred to watch a light rather than listen to a telephone. The speaking galvanometer used is a specially constructed one, and does not easily get out of order, so that, everything being once arranged, the men had only to keep the lamp in order.

Messrs Smith & Granville devised a novel and strong form of apparatus for a "call," and by its means any number of bells could be rung, thus securing attention. The instruments both on rock and shore were identical, and, in actual work, two to three Leclanché cells were ample.

By the means above described, communication was obtained through the gap of water 60 yards in length. This by no means is the limit, for it will be apparent that the gap distance is determined by the volume of water in the immediate neighbourhood of the rock, as well as by the sensitiveness of the receiving instrument and the magnitude of the sending current.

This method is well suited for coast defences. For instance, if a cable is laid from the shore out to sea, with its end anchored in a known position, then it would be easy for any ship, knowing the position of the submerged end, to communicate with shore by simply lowering (within one or two hundred yards of the anchored end) an insulated wire having the end of its conductor attached to a small mass of metal to serve as "earth," the circuit being completed through the hull of the ship and the sea.[1]

As this method has been in practical use at the Fastnet Lighthouse for the last three years, the following account of the installation, which has been kindly supplied by Mr W. S. Smith, will be of interest:—

"The difficulty of maintaining electrical communication with outlying rock lighthouses is so great that it has become necessary to forego the advantages naturally attendant upon the use of a submarine cable laid in the ordinary way continuously from the shore to the lighthouse, inasmuch as that portion of the cable which is carried up from the sea-bed to the rock is rapidly worn or chafed through by the combined action of storm and tide. By the use of the Willoughby Smith & Granville system of communication this difficulty is avoided, for the end of the cable is not landed on the rock at all, but terminates in close proximity thereto and in fairly deep undisturbed water. This system, first suggested in 1887 and practically demonstrated at the Needles Lighthouse in 1892, has—on the recommendation of the Royal Commission on Lighthouse and Lightship Communication — been applied to the Fastnet, one of the most exposed and inaccessible rock lighthouses of the United Kingdom.

"The Fastnet Rock, situated off the extreme S.W. corner of Ireland, is 80 feet in height and 360 feet in length, with a maximum width of 150 feet, and is by this system placed

[1] 'Electrician,' September 29, 1893. See also the 'Times,' November 24, 1892.

in electrical communication with the town of Crookhaven, eight miles distant.

"The shore end of the main cable, which is of ordinary construction, is landed at a small bay called Galley Cove, about one mile to the west of the Crookhaven Post Office, to which it is connected by means of a subterranean cable of similar construction having a copper conductor weighing 107 lb. covered with 150 lb. of gutta-percha per nautical mile. The distant or sea end of the main cable terminates seven miles from shore, in 11 fathoms of water, at a spot about 100 feet from the Fastnet Rock; and the end is securely fastened to a copper mushroom-shaped anchor weighing about 5 cwt., which has the double duty of serving electrically as an 'earth' for the conductor, and mechanically as a secure anchor for the cable end.

"The iron sheathing of the last 100 feet of the main cable is dispensed with, so as to prevent the possibility of any electrical disturbance being caused by the iron coming in contact with the copper of the mushroom; and, as a substitute, the conductor has been thickly covered with india-rubber, then sheathed with large copper wires, and again covered with india-rubber—the whole being further protected by massive rings of toughened glass.

"To complete the main cable circuit, a short earth line, about 200 yards in length, is laid from the post office into the haven.

"By reference to the diagram (fig. 29) it will be seen that if a battery be placed at the post office, or anywhere in the main cable circuit, the sea becomes electrically charged —the charge being at a maximum in the immediate vicinity of the mushroom, and also at the haven 'earth.' Under these conditions, if one end of a second circuit is inserted in the water anywhere near the submerged mushroom—for instance, on the north side of the Fastnet—it partakes, more or less, of the charge; and if the other end of this second circuit is also connected to the water, but at a point more remote from the mushroom—for instance, at the south side of the Fastnet—then a current will flow in the second circuit, due to the difference in the degree of charge at the two ends; and accordingly a galvanometer or other sensitive

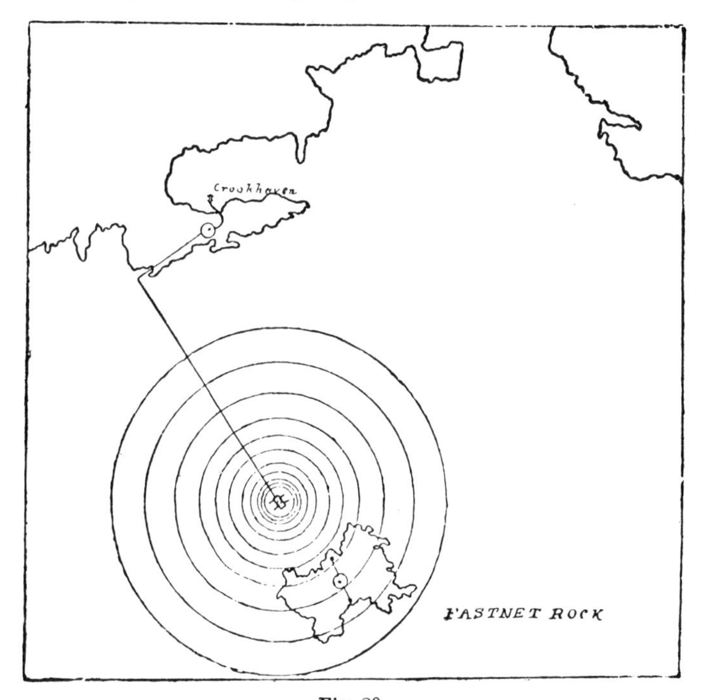

Fig. 29.

instrument placed in the Fastnet circuit is affected whenever the post office battery is inserted in the main cable circuit, or, *vice versâ*, a battery placed in the Fastnet circuit will affect a galvanometer at the post office.

"In practice ten large-size Leclanché cells are used on the rock, the sending current being about 1·5 ampères, and in this case the current received on shore is equal to about ·15 of a milliampère. The received current being small, instruments of a fair degree of sensitiveness are required, and such instruments, when used in connection with cables having both ends direct to earth, are liable to be adversely affected by what are known as 'earth' and 'polarisation' currents, consequently special means have been devised to prevent this.

"The receiving instrument is a D'Arsonval reflecting galvanometer, which has been modified to meet the requirements by mounting the apparatus on a vertical pivot, so that by means of a handle the galvanometer can be rotated through a portion of a circle—thus enabling the zero of the instrument to be rapidly corrected. This facility of adjustment is necessary on account of the varying 'earth' and 'polarisation' currents above mentioned.

"An entirely novel and substantial 'call' apparatus has also been designed, which automatically adapts itself to any variation in the earth or polarisation current. It consists essentially of two coils moving in a magnetic field, and these coils are mounted one at each end of a balanced arm suspended at its centre and free to rotate horizontally within fixed limits. The normal position of the arm is midway between two fixed limiting stops. Any current circulating in the coils causes the whole suspended system to rotate until the arm is brought into contact with one or other of the stops—the direction of rotation depending upon the direction of the current. A local circuit is thus closed, which releases a clockwork train connected to a torsion head carrying the suspending wire, and thus a counterbalancing twist or torsion is put into the wire, and this torsion slowly increases until the arm leaves the stop and again assumes its free position. If, however, the current is reversed within a period of say five or ten seconds, then the clockwork closes a second circuit and the electric bell is operated. By this arrangement, whilst the relay automatically adjusts itself for all variations of current, the call-bell will only respond to definite reversals of small period and not to the more sluggish movements of earth-currents. It is evident that one or more bells can be placed in any part of the building. The receiving galvanometer and the 'call' relay have worked very satisfactorily, and any man of average intelligence can readily be taught in two or three weeks to work the whole system.

"To enable the two short cables that connect the lighthouse instruments with the water to successfully withstand the heavy seas that at times sweep entirely over the Fastnet, it has been found necessary to cut a deep 'chase' or groove down the north and south faces of the rock from summit to near the water's edge, and to bed the cables therein by means of Portland cement. And since the conductors must make connection with the water at all states of sea and tide, two slanting holes 2½ inches in diameter have been drilled through the solid rock from a little above low-water mark to over 20 feet below. Stout copper rods connected with the short cables are fitted into these holes, and serve to maintain connection with the water even in the roughest weather, and yet are absolutely protected from damage."

Mr Granville supplies some interesting particulars as to the difficulties of their installation at the Fastnet.[1] "The rock," he says, "is always surrounded with a belt of foam, and no landing can be made except by means of a jib 58 feet long—not at all a pleasant proceeding. Now, here is a case where the Government desired to effect communication telegraphically, but, as had been proved by very costly experiments, it was impossible to maintain a continuous cable, the cable being repeatedly broken in the immediate vicinity

[1] 'Jour. Inst. Elec. Engs.,' No. 137, p. 941.

of the rock. This, therefore, is a case where some system of wireless telegraphy is absolutely necessary, but neither of the systems described would answer here.[1] Dr Lodge advises us to eschew iron, and to avoid all conducting masses. But the tower and all the buildings are built of boiler-plate, and that which is not of iron is of bronze. In fact, the rock itself is the only bit of non-conducting, and therefore non-absorbing, substance for miles around. It is very clear in a case of this sort—and this is a typical case—that it is absolutely impracticable to employ here Dr Lodge's method. Now we hear in regard to the method used—and successfully used—at Lavernock, that a certain base is required, of perhaps half a mile, a quarter of a mile, or a mile in length; and that base must bear some proportion to the distance to be bridged. But where can you get any such base on the rock? You could barely get a base of 20 yards, so *that* method utterly fails. Then we come to the case suggested by Mr Evershed, of a coil which would be submerged round the rock. Well, where would the coil be after the first summer's breeze, let alone after a winter gale? Why, probably thrown up, entangled, on the rock. A few years ago, during a severe gale, the glass of the lantern, 150 feet above sea-level, was smashed in; and at the top of the rock, 80 feet above the sea-level, the men dare not, during a winter's gale, leave the shelter of the hut for a moment, for, as they said,—and I can well believe it,—they would be swept off like flies. This is a practical point, and therefore one I am glad to bring to the notice of the Institution; and, I repeat, if wireless telegraphy is to be of use, it must be of use for these exceptional cases."

Strange as it may seem, we have been using, on occasion, wireless telegraphy of this form for very many years without

[1] *I.e.*, those advocated by Professor Lodge and Mr Sydney Evershed. See 'Jour. Inst. Elec. Engs.,' No. 137, pp. 799, 852.

recognising the fact. Every time in ordinary telegraphy that we "work through a break," as telegraphists say, we are doing it. An early instance of the kind is described in the old 'Electrician,' January 9 and 23, 1863. Many years ago, in Persia, the author has often worked with the ordinary Morse apparatus through breaks where the wire has been broken in one or more places, with the ends lying many yards apart on damp ground, or buried in snow-drifts. As the result of his experiences in such cases the following departmental order was issued by the Director, Persian Telegraphs, as far back as November 2, 1881: "In cases of total interruption of all wires, it is believed that communication may in most cases be kept up by means of telephones. Please issue following instructions: Fifteen minutes after the disappearance of the corresponding station, join all three wires to one instrument at the commutator. Disconnect the relay wire from the key of said instrument, and in its stead connect one side of telephone, other side of which is put to earth. Now call corresponding station slowly by key, listening at telephone for reply after each call. Should no reply be received, or should signals be too weak, try each wire separately, and combined with another, until an arrangement is arrived at which will give the best signals." The Cardew sounder or buzzer has in recent years been added, and with very good results. It will thus be seen that Mr Willoughby Smith's plan is really an old friend in a new guise.

In 1896 Mr A. C. Brown, of whose work in wireless telegraphy we have already spoken (p. 101, *supra*), revived the early proposals of Gauss (p. 3), Lindsay (p. 20), Highton (p. 40), and Dering (p. 48), *re* the use of bare wire, or badly insulated cables, in connection with interrupters and telephones. He also applied his method to cases where the continuity of the cable is broken. "Providing the ends remain anywhere in proximity under the water, communication can usually be kept up, the telephone receivers used in this way being so exceedingly sensitive that they will respond to the very minute traces of current picked up by the broken end on the receiving side from that which is spreading out through the water in all directions from the broken end on the sending side." (See Mr Brown's patent specification, No. 30,123, of December 31, 1896.)

Recently he has been successful in bridging over in this way a gap in one of the Atlantic cables; but in this he has done nothing more than the present writer did in 1881, and Mr Willoughby Smith in 1887.

G. MARCONI'S METHOD.

> "Even the lightning-elf, who rives the oak
> And barbs the tempest, shall bow to that yoke,
> And be its messenger to run."
>
> —*Supple's Dampier's Dream.*

We now come to the crowning work of Mr Marconi in wireless telegraphy; but before describing this method it will be desirable to make ourselves acquainted with the principles involved in the special apparatus which he employs, and which differentiates his system from all those that have hitherto occupied us. For this we need only go back a few years, and make a rapid survey of the epoch-marking discoveries of a young German philosopher, Heinrich Hertz.[1]

To properly appreciate the work of Hertz we must carry

[1] Hertz was born in Hamburg, February 22, 1857, and died in Bonn, January 1, 1894. For interesting notices of his all too brief life, see, *inter alia*, the 'Electrician,' vol. xxxiii. pp. 272, 299, 332, and 415.

our minds back two hundred years, to the time when Newton made known to the world the law of universal gravitation. Here, in the struggle between Newtonianism and the dying Cartesian doctrine, we have the battle-royal between the rival theories of action-at-a-distance and action-by-contact. The victory was to the former for a time; and in the hands of Bernouilli, and, subsequently, of Boscovich, the doctrines of Newtonianism were carried far beyond the doctrines of the individual Newton. In fact, Newton expressed himself as being opposed to the notion of matter acting where it is not; though, as we see by his support of the emission theory of light, he was not prepared to accept the notion of a luminiferous ether. Newton, however, suggested that gravitation might be explained as being due to a diminution of pressure in a fluid filling space. Thus the doctrine of an empty space, requiring the infinitely rapid propagation of a distance-action, held the field, and was recognised by scientists of the eighteenth century as the only plausible hypothesis.

History repeats itself; and again the battle-royal was fought, this time, early in the nineteenth century, in favour of the ether hypothesis; and action-at-a-distance was mortally wounded. Before the phenomena of interference of light and the magnetic and electro-static researches of Faraday, both the idea of empty space action and that of the emission of light failed; and the propagation of force through the ether, and of light by vibratory conditions of the ether, came to be held as necessary doctrines. Later still,[1] Maxwell assumed the existence of, and investigated the state of, stress in a medium through which electromagnetic action is propagated. The mathematical theory

[1] October 1864, in his paper on the Dynamical Theory of the Electro-Magnetic Field, 'Phil. Trans.,' vol. 155. See also his great work, 'Electricity and Magnetism,' published in 1873.

which he deduced gives a set of equations which are identical in form with the equations of motion of an infinite elastic solid; and, on this theory, the rate of propagation of a disturbance is equal to the ratio of the electro-magnetic and electro-static units. The experimental determination by Maxwell and others, that this ratio is a number equal to the velocity of light in ether in centimetres per second, is a fact which gave immense strength to the Maxwellian hypothesis of identity of the light and electro-magnetic media. But, although this is the case, the Maxwellian hypothesis, even when taken in conjunction with the experimental support which he educed for it, fell far short of being a complete demonstration of the identity of luminous and electro-magnetic propagation.[1]

To the genius of Hertz we owe this demonstration. One of the most important consequences of Maxwell's theory was that disturbances of electrical equilibrium produced at any place must be propagated as waves through space, with a velocity equal to that of light. If this propagation was to be traced through the small space inside a laboratory, the disturbances must be rapid, and if a definite effect was to be observed, they must follow each other at regular intervals; in other words, periodical disturbances or oscillations of extreme rapidity must be set up, so that the corresponding wave-length, taking into account the extraordinarily high velocity of propagation (186,000 miles per second), may be only a few inches, or at most feet. Hertz was led to an experiment which satisfied these conditions, and thus supplied the experimental proof which Maxwell and his school knew must come sooner or later.

The oscillatory nature of the discharge of a Leyden jar, under certain conditions, was theoretically deduced by Von Helmholtz in 1847; its mathematical demonstration was

[1] Lord Kelvin's Address, Royal Society, November 30, 1893.

given by Lord Kelvin in 1853; and it was experimentally verified by Feddersen in 1859. When a Leyden jar, or a condenser, of an inductive capacity K, is discharged through a circuit of resistance R and self-induction L, the result is an instantaneous flow, or a series of oscillations, according as R is greater, or less, than $2\sqrt{\frac{L}{K}}$; and in the latter case the oscillatory period or amplitude is given in the equation—

$$T = 2\pi\sqrt{KL}$$

where π is the constant 3·1415 ('Phil. Trans.,' June 1853).[1]

In his collected papers[2] Hertz tells us that his interest in the study of electrical oscillations was originally awakened by the announcement of the Berlin prize of 1879, which was to be awarded for an experimental proof of a relation between electro-dynamic forces and dielectric polarisation in insulators. At the suggestion of his master and friend, Von Helmholtz, the young philosopher took up the inquiry, but soon discovered that the then known oscillations were too slow to offer any promise of success, and he gave up the immediate research; but from that time he was always on the look-out for phenomena in any way connected with the subject. Consequently, he immediately recognised the importance of a casual observation which in itself and at another time might have been considered as too trivial for further notice. In the collection of physical apparatus at Karlsruhe he found an old pair of so-called Riess's or Knochenhauer's spirals—short flat coils of insulated wire,

[1] In the old 'Electrician,' vol. iii. p. 101, there is an interesting paper on "The Oscillatory Character of Spark Discharges shown by Photography." For a concise exposition of the theory of electrical oscillations, see Prof. Edser's paper, 'Electrical Engineer,' June 3, 1898, and following numbers.

[2] 'Electric Waves,' London, 1893. For an interesting account of pre-Hertzian observations, see Lodge's 'The Work of Hertz,' p. 61; also Appendix D of this work.

with the turns all in the same plane (? Prof. Henry's spirals). While performing some experiments with them at a lecture he was giving, he noticed that the discharge of a very small Leyden jar, or of a small induction coil, passed through the one was able to excite induced currents in the other, provided that a small spark-gap was made in the circuit of the first spiral. Thus was made the all-important discovery of the "effective spark-gap" which started Hertz on the road of his marvellous investigations.

A very little consideration of this phenomenon enabled him, even at this early stage, to lay down the following propositions :—

1. If we allow a condenser, such as a Leyden jar, of small capacity, to discharge through a short and simple circuit with a spark-gap of suitable length, we obtain a sharply defined discharge of very short duration, which is the long-sought-for sudden disturbance of electrical equilibrium—the *exciter* of electrical vibrations.

2. Such vibrations are capable of exciting in another circuit of like form resonance effects of such intensity as to be evident even when the two circuits are separated by considerable distances. In this second circuit Hertz had found the long-sought-for *detector* of electric waves.

With the exciter to originate electric waves and the detector to make them evident at a distance, all the phenomena of light were, one after another, reproduced in corresponding electro-magnetic effects, and the identity of light and electricity was completely demonstrated.[1]

In his paper "On very Rapid Electric Oscillations," Hertz occupied himself with some of these phenomena. As an exciter he used wire rectangles, or simple rods (fig. 30) to the ends of which metallic cylinders or spheres were con-

[1] See Appendix A for a clear exposition of the views regarding the relation of the two *before* and *after* Hertz.

nected, the continuity being broken in the middle where the ends were provided with small spherical knobs between

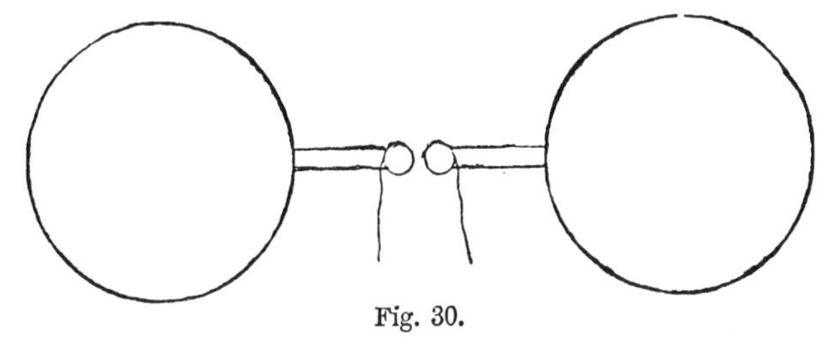

Fig. 30.

which the sparks passed. The exciter was charged by an ordinary Ruhmkorff induction coil of small size.

The detector was mostly a simple rectangle or circle of wire (fig. 31), also provided with a spark-gap. When vibrations are set up in the detector and sparks pass across the gap, the greater length of these sparks indicates the greater intensity of the received wave impacts. When, therefore, the dimensions of the detector are so adjusted as to give the maximum sparks with a given exciter the two circuits are said to be in resonance, or to be electrically tuned. Fortunately this condition of resonance or syntony is not essential to the excitement of sparks, else wireless telegraphy by Hertzian waves would not be so advanced as it is to-day. Thus, when a good exciter is in action it will cause little sparks between any conducting body in its vicinity and a wire held in the hand and brought near to the body, showing that the influence of the exciter extends to all conducting bodies, and not merely to those which are tuned to it. Of course it still holds good that, *cœteris paribus*, the maximum effect is obtained with resonance.

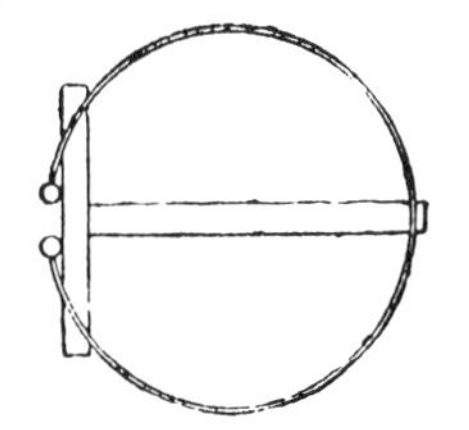

Fig. 31.

In the course of his experiments on electric resonance, Hertz observed a phenomenon which for a time was inexplicable. It was seen that the length and brightness of the sparks at the detector were greatly modified by the sparks given off at the exciter. If the latter were visible from the detector spark-gap the sparks given off there were small and hardly perceptible, but became larger and brighter as soon as a screen was placed between the two instruments. By carefully thought-out experiments he showed that this singular action was due solely to the presence of ultra-violet light, breaking down the insulation of the gap and making it, so to say, more conductive. This effect can be shown in another way, by widening the spark-gap of an induction coil beyond the ordinary sparking distance, when, by simply directing a beam of ultra-violet light into the gap, sparking will be resumed.[1]

Having made himself familiar with the phenomena of electrical resonance, Hertz went on to study the propagation of electric vibrations through space—the most difficult, as it is probably the most important, of all his researches.

[1] Prof. K. Zickler has proposed to use this property for telegraphy. At the sending station an arc lamp, which is rich in ultra-violet rays, is provided with a shutter and a lens for directing flashes towards the receiving station. There they are made to impinge on the spark-gap, unduly widened, of an induction coil in action, and allow sparks to pass. These give rise to electric waves which act on the coherer, which in its turn operates a bell, a telephone, or a Morse instrument in the way we shall see later on when we come to speak of the action of the Marconi apparatus. The reflecting lens is made of quartz and not of glass, which does not transmit the ultra-violet rays ; but for signalling or interrupting the rays in long and short periods a glass plate is used as the shutter. The interruption of the ultra-violet rays is thus effected without altering the light, which assures secrecy of transmission. Prof. Zickler has in this way signalled over a space of 200 metres, and thinks that with suitable lamps and reflectors the effect would be possible over distances of many kilometres.—'Elektrische Zeitung,' July 1898.

The results he gave to the world in 1888, in his paper "On the Action of a Rectilinear Electric Oscillation on a Neighbouring Circuit." When sparks pass rapidly at the exciter electric surgings occur, and we have a rectilinear oscillation which radiates out into surrounding space. The detectors, whose spark-gaps were adjustable by means of a micrometer screw, were brought into all kinds of positions with respect to the exciter, and the effects were studied and measured. These effects were very different at different points and in the different positions of the detector. In short, they were found to obey a law of radiation which was none other than the corresponding law in optics.

In his paper, "On the Velocity of Propagation of Electro-dynamic Actions," he gave experimental proof of the hitherto theoretical fact that the velocity of electric waves in air was the same as that of light, whereas he found the velocity in *wires* to be much smaller—in the ratio of 4 to 7. For the moment he was puzzled by this result: he suspected an error in the calculations, or in the conditions of the experiment, but—and here he showed himself the true philosopher—he did not hesitate to publish the actual results, trusting to the future to correct or explain the discrepancy. The explanation was soon forthcoming. Messrs E. Sarasin and L. de la Rive of Geneva took up the puzzle, and ended by showing that the deviations from theory were caused simply by the walls of Hertz's laboratory, which reflected the electric waves impinging on them, so causing interferences in the observations. When these investigators repeated the Hertzian experiment with larger apparatus, and on a larger scale, as they were able to do in the large turbine hall of the Geneva Waterworks, they found the rate of propagation to be the same along wires as in air.[1]

[1] 'Comptes Rendus,' March 31, 1891, and December 26, 1892. See also the 'Electrician,' vol. xxvi. p. 701, and vol. xxx. p. 270.

In his paper, "On Electro-dynamic Waves and their Reflection," Hertz further developed this point, and showed the existence of these waves in free space. Opposite the exciter a large screen of zinc plate, 8 feet square, was suspended on the wall; the electric waves emitted from the exciter were reflected from the plate, and on meeting the direct waves interference phenomena were produced, consisting of stationary waves with nodes and loops. When, therefore, Hertz moved the circle of wire which served as a detector to and fro between the screen and the exciter, the sparks in the detector circuit disappeared at certain points, reappeared at other points, disappeared again, and so on. Thus there was found a periodically alternating effect corresponding to nodes and loops of electric radiation, showing clearly that in this case also the radiation was of an undulatory character, and the velocity of its propagation finite.

In a paper, "On the Propagation of Electric Waves along Wires," March 1889, Hertz shows that alternating currents or oscillations of very high frequencies, as one hundred million per second, are confined to the surface of the conductor along which they are propagated, and do not penetrate the mass.[1] This is a very important experimental proof of Poynting's theory concerning electric currents, which he had deduced from the work of Faraday and Maxwell. According to this theory, the electric force which we call the current is in nowise produced *in* the wire, but under all circumstances enters from without, and spreads itself in the metal comparatively slowly, and according to similar laws as

[1] It should be stated here that long ago Prof. Henry, the Faraday of America, held the same views, and proved them, too, by an experiment which is strangely like one of Hertz's, though, of course, he did not explain them as Hertz does. Henry's views are given clearly in two letters addressed to Prof. Kedzie of Lansing, Michigan, in 1876. Being of historical interest, as well as of practical value, I give them entire in Appendix B.

govern changes of temperature in a conductor of heat. If the electric force outside the wire is very rapidly altering in direction or oscillating, the effect will only enter to a small depth *in* the wire ; the slower the alterations occur, the deeper will the effect penetrate, until finally, when the changes follow one another infinitely slowly, the electric effect occupies the whole mass of the wire with uniform density, giving us the phenomenon of the so-called current.

In support of this view Hertz devised many beautiful experiments, one or two of which may be described here.

If a primary conductor acts through space upon a secondary conductor, it cannot be doubted that the effect reaches the latter from without. For it can be regarded as established that the effect is propagated in space from point to point, therefore it will be forced to meet first of all the outer boundary of the body before it can act upon the interior of it. But a closed metallic envelope is shown to be quite opaque to this effect. If we place the secondary conductor in such a favourable position near the primary one that we obtain sparks 5 to 6 millimetres long, and then surround it with a closed box made of zinc plate, the smallest trace of sparking can no longer be perceived. The sparks similarly vanish if we entirely surround the primary conductor with a metallic box. It is well known that with relatively slow variations of current the integral force of induction is in no way altered by a metallic screen. This is, at the first glance, contradictory to the present contention. However, the contradiction is only an apparent one, and is explained by considering the duration of the effects. In a similar manner a screen which conducts heat badly protects its interior completely from rapid changes of the outside temperature, less from slow changes, and not at all from a continuous rising or lowering of the temperature. The thinner the screen is, the more are the rapid variations of the outside temperature felt in its interior.

In our case also the electrical action must plainly penetrate into the interior of the closed box, if we only diminish sufficiently the thickness of the metal. But Hertz did not succeed in attaining the necessary thinness, —a box covered with tinfoil protected completely, and even a box of gilt paper, if care was taken that the edges of the separate pieces of paper were in metallic contact. In this case the thickness of the conducting metal was estimated to be barely $\frac{1}{20}$ millimetre. To demonstrate this, he fitted the protecting envelope as closely as possible round the secondary conductor, and widened the spark-gap to about 20 millimetres, adding an auxiliary spark-gap exactly opposite to it. The sparks were in this case not so long as in the ordinary arrangement, since the effect of resonance was now wanting, but they were still very brilliant. Between the ends of this envelope, then, brilliant sparks were produced; but on observing the auxiliary spark-gap (through a wire-gauze window in the envelope), not the slightest electrical movement could be detected in the interior.

The result of the experiment is not affected if the envelope touches the conductor at a few points: the insulation of the two from each other is not necessary in order to make the experiment succeed, but only to give it the force of a proof. Clearly we can imagine the envelope to be drawn more closely round the conductor than is possible in the experiment; indeed, we can imagine it to coincide with the outermost layer of the conductor. Although, then, the electrical disturbances on the surface of our conductor are so powerful that they give sparks 5 to 6 millimetres long, yet at $\frac{1}{20}$ millimetre beneath the surface there exists such perfect freedom from disturbance that it is not possible to obtain the smallest sparks. We are brought, therefore, to the conclusion that what we call an induced current in the secondary conductor is a phenomenon which is manifested in its neighbourhood, but to which its interior scarcely contributes.

One might grant that this is the state of affairs when the electric disturbance is conveyed through a dielectric, but maintain that it is another thing if the disturbance, as one usually says, has been propagated in a conductor. Let us place near one of the end plates of our primary conductor a conducting-plate, and fasten to it a long, straight wire: we have already seen (in previous experiments) how the effect of the primary oscillation can be conveyed to great distances by the help of this wire. The usual theory is that a wave travels along the wire. But we shall try to show that all the alterations are confined to the space outside and the surface of the wire, and that its interior knows nothing of the wave passing over it.

Hertz arranged experiments first of all in the following manner: A piece about 4 metres long was removed from the wire conductor and replaced by two strips of zinc plate 4 metres long and 10 centimetres broad, which were laid flat one above the other, with their ends permanently connected together. Between the strips along their middle line, and therefore almost entirely surrounded by their metal, was laid along the whole 4 metres' length a copper wire covered with gutta-percha. It was immaterial for the experiments whether the outer ends of this wire were in metallic connection with, or insulated from, the strips: however, the ends were mostly soldered to the zinc strips. The copper wire was cut through in the middle, and its ends were carried, twisted round each other, outside the space between the strips to a fine spark-gap, which permitted the detection of any electrical disturbance taking place in the wire. When waves of the greatest possible intensity were sent through the whole arrangement there was nevertheless not the slightest effect observable in the spark-gap. But if the copper wire was displaced anywhere a few decimetres from its position, so that it projected just a little beyond the space between the strips, sparks immediately began to pass. The sparks were the more intense according to the length of copper wire extending beyond the edge of the zinc strips and the distance it projected. The unfavourable relation of the resistances was therefore not the cause of the previous absence of sparking, for this relation had not been changed; but the wire being in the interior of the conducting mass was at first deprived of the influence coming from without. Moreover, it is only necessary to surround the projecting part of the wire with a little tinfoil in metallic communication with the zinc strips, in order to immediately stop the sparking again. By this means we bring the copper wire back again into the interior of the conductor.

We can conclude, then, that rapid electric oscillations are unable to penetrate metallic sheets or wires of any thickness, and that it is, therefore, impossible to produce sparks by the aid of such oscillations in the interior of closed metallic screens. If, then, we see sparks so produced in the interior of metallic envelopes which are nearly, but not quite, closed, we must conclude that the electric disturbance has forced itself in through the openings. Let us take a typical case of this kind.

In fig. 31A we have a wire cage *A* just large enough to hold the spark-gap. One of the discs *a* is in metallic connection with the central wire; the other *b* is clear of the wire (which passes freely through the central hole), but is connected to the metallic tube *c*, which completely surrounds (without touching it) the central wire for a length of 1·5 metre. On sending a series of waves through this arrangement in the direction shown by the arrow, we obtain brilliant sparks at *A*, which do not become materially smaller, if, without making any other alteration, we lengthen the tube *c* to as much as 4 metres.

According to the old theory, it would be said that the wave arriving at *A* penetrates easily the thin metallic disc *a*, leaps across the spark-gap, and travels on in the central wire; but according to the present view, the explanation is as follows: The wave arriving at *A* is quite unable to penetrate the disc *a*; it therefore glides over it, over the outside of the apparatus, and on to the point *d*, 4 metres distant. Here it divides: one part travels on along the wire; the other bends into the interior of the tube, and runs

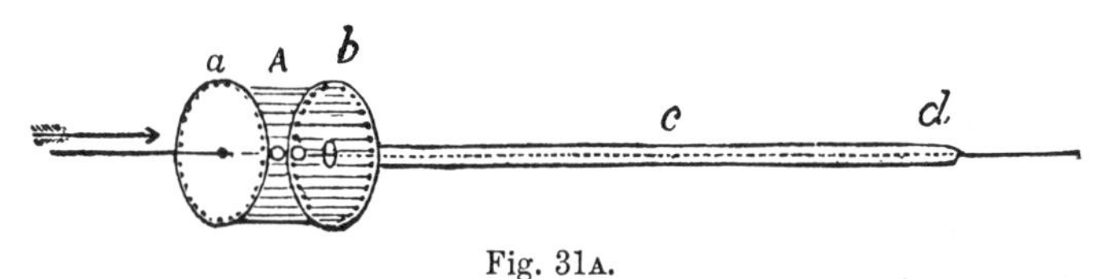

Fig. 31A.

back in the space between the tube and the wire to the spark-gap, where it gives rise to the sparking. That this view is the correct one is shown by the fact, amongst others, that every trace of sparking disappears as soon as we close the opening at *d* by a tinfoil stopper.

Reviewing his experiments on this subject, Hertz says: "A difference will be noticed between the views here put forward and the usual theory. According to the latter, conductors are represented as those bodies which alone take part in the propagation of electric disturbances; non-conductors are the bodies which oppose this propagation. According to our view, on the contrary, all transmission of electrical disturbances is brought about by non-conductors; conductors oppose a great resistance to any rapid changes in this transmission. One might almost be inclined to maintain that conductors and non-conductors should, on this theory, have their names interchanged. However, such a paradox only arises because one does not specify the kind of conduction or non-conduction considered. Undoubtedly metals are non-conductors of electric force, and just for this reason they compel it under certain circumstances to remain concentrated instead of becoming dissipated; and thus they become conductors of the apparent source of these forces, electricity, to which the usual terminology has reference."[1]

In the course of his experiments Hertz had succeeded in producing very short electric waves of 30 centimetres in length, the oscillations corresponding to which could be collected by a concave cylindrical mirror and concentrated into a single beam of electric radiation. According to Maxwell's theory of light, such a beam must behave like a beam of light, and that this is the case Hertz abundantly proved in his next paper, "On Electric Radiation." He showed how such radiation was propagated in straight lines like light; that it could not pass through metals, but was reflected by them; that, on the other hand, it was able to penetrate wooden doors and stone walls. He also proved, by setting up metallic screens, that a space existed behind them in which no electric action could be detected, thus producing electric shadows; and, by passing the electric rays through a wire grating, he was able to polarise them, just as light is polarised by passage through a Nicol prism.

[1] As this is a matter of some complexity to all who, like myself belong to the old way of thinking—the *ancien régime*—and as, moreover, it is of great practical importance, especially as regards the proper construction of lightning protectors, and the supply mains of electric light and power, I have thought it useful to give in Appendix B some extracts, which I hope will make the new views intelligible to the ordinary reader. Lodge's 'Modern Views of Electricity' should also be consulted.

Perhaps the most striking experiment of all in this field was his last one, in which he directed the ray on to a large pitch prism weighing 12 cwts. : the ray was deflected, being, in fact, refracted like a ray of light in a glass prism.

Thus he gave to the experimental demonstration of Maxwell's electro-magnetic theory of light its finishing touch, and the edifice was now complete. Hertz's marvellous researches were presented in succession, as rapid and surprising almost as the sparks with which he dealt, to the Berlin Academy of Sciences, between November 10, 1887, and December 13, 1889. They were collected and published in book form, in 1893, under the title of 'Electric Waves' (English translation edited by Prof. D. E. Jones), to which the reader is referred for further information.[1]

Here it will suffice, in conclusion, to briefly sum up the chief results of these epoch-making investigations. In the first place, Hertz has freed us from the bondage of the old theory of action-at-a-distance; and as regards electric and magnetic effects, he has shown that they are propagated through the ether which fills all space and with finite velocity. The mysterious darkness which surrounded those strange distance-actions—that something can act where it is not—has now been cleared away. Further, the identity of the form of energy in the case of two powerful agents in nature has been conclusively established; light and electrical radiation are essentially the same, different manifestations of the same processes, and so the old elastic-solid theory of optics is resolved into an electro-magnetic theory. The velocity of propagation of light is the same as that of electro-magnetic waves, and these in turn obey all the laws of optics. The scope of optics is thus enormously widened; to the ultra-violet, visible, and infra-red rays, with their wave-lengths

[1] Our account of Hertz's investigations is chiefly drawn from Prof. Ebert's paper in the 'Electrician,' vol. xxxiii. pp. 333-335.

of thousandths of a millimetre, are now to be added, lower down the scale, electro-magnetic waves, producible in any length from fractions of an inch to thousands of miles.

Hertz's ordinary waves were many metres long, and he does not appear to have ever worked with waves of less than 30 centimetres. Righi, however, by employing exciters with small spheres, obtained waves of 2·5 centimetres; while Prof. Chunder Bose of Calcutta, using little pellets of platinum, was able to produce them of only 6 millimetres! The smaller the exciter and its pellets the shorter the waves, until we come in imagination to the exciter—the ultimate molecule, whose waves should approximate to light.

The following table compares approximately some of the known vibrations in ether and air :—

Ether vibrations per second—

billions (?)	=	Röntgen rays.
10,000 " (?)	=	Actinic "
8,000 "	=	Violet "
5,500 "	=	Green "
4,000 "	=	Red "
2,800 "	=	Infra red "
1,000 to 2,000 "	=	Radiant heat.
50 thousands to 2,000 billions	=	Hertzian waves.

Air vibrations per second—

33,000	=	Highest audible note.
4,000	=	Highest musical note.
2,000	=	Highest soprano.
150 to 500	=	Ordinary voice.
32	=	Lowest musical note.
16	=	" audible "

The work of Hertz was immediately taken up, and is now being carried on (doubtless towards fresh conquests, for there is no finality in science) by a whole army of investigators, of whom we need only mention a few—as Lodge, Righi, Branly, Sarasin, and de la Rive—whose discoveries, especially as regards the exciter and detector, more immediately concern us in this history.

The exciter of Hertz, although sufficing for his special purposes, had the disadvantage that the sparks in a short time oxidised the little knobs and roughened their surfaces, which made their action irregular and necessitated their frequent polishing. Messrs Sarasin and de la Rive of Geneva obviated this difficulty by placing the knobs in a vessel containing olive-oil. The effect of this arrangement was at once to augment the sparks at the detector, so that when it was placed close to the exciter the sparks were a perfect blaze; and at 10 metres' distance, with detectors of large diameter (75 to 1 metre), they were still very bright and visible from afar. It is true that here, too, the oil carbonises in time and loses its transparency; but if a considerable quantity, as two or three litres, be employed, there is no perceptible heating, and the intensity of the sparks is hardly altered, even after half an hour's continuous working. Prof. Righi substituted vaseline-oil, made suitably thick by the addition of solid vaseline. His exciter is composed of two metal balls, each set in an ebonite frame; a parchment envelope connects these frames and contains the oil which thus fills the spark-gap. Righi attributes the increased efficiency of his exciter (1) to the heightening effect which a cushion of (insulating) liquid seems to have on the electric potential which gives rise to the sparks—a sort of (to adopt an expressive French phrase) *reculant pour mieux sauter;* and (2) to some sort of regularising effect making their production more uniform. Like Sarasin and de la Rive, he found that the use of vaseline obviated the necessity of frequent cleaning of the knobs, for even after long usage, when the liquid had become black and a deposit of carbon had formed on the opposing surfaces, the apparatus continued to work satisfactorily. Righi also found that solid knobs gave better results than hollow ones, the oscillations in the former case being perceptible in the detector at nearly double the distance attained in the latter case.

The detector usually employed by Hertz consisted of a metal wire bent into a rectangle or a circle (see fig. 31), and terminated by two little knobs between which the sparks played. But this form is not obligatory: any two distinct conducting surfaces separated by a spark-gap will serve equally well. Many kinds of detectors have been employed, but in this place we need only concern ourselves with those of the microphonic order, which alone enter into the construction of the Marconi system of telegraphy.[1]

Just mentioning the well-known electrical behaviour of selenium under the action of light; the fact observed by Prof. Minchin that his delicate "impulsion-cells" were affected by Hertzian waves; the Righi detector, consisting of thin bands of quicksilver (as used for mirrors) rendered discontinuous by cross-lines lightly traced with a diamond; and the original Lodge "coherer," consisting of a metallic point lightly resting on a metal plate,[2]—we come to the special form known as Branly's detector, or, as he prefers to call it, the radio-conductor.

The observance of the phenomena underlying Branly's detector goes back further than is usually supposed. Thus, Mr S. A. Varley, as long ago as 1866, noticed some of them,

[1] For other forms of detectors, based on physiological, chemical, electrical, thermal, and mechanical principles, see Lodge's 'The Work of Hertz and his Successors,' pp. 25, 56.

[2] For the first suggestions of Lodge's detector see his paper, "On Lightning-Guards for Telegraphic Purposes," 'Jour. Inst. Elec. Engs.,' vol. xix. pp. 352-354. Even before this the learned professor succeeded in detecting electric waves by means of a telephone, 'Jour. Inst. Elec. Engs.,' vol. xviii. p. 405.

and applied them in the construction of a lightning protector for telegraph apparatus.

In his paper read before the British Association (Liverpool meeting, 1870), he says :—

"The author, when experimenting with electric currents of varying degrees of tension, had observed the very great resistance which a loose mass of dust composed even of conducting matter will oppose to electric currents of moderate tension.

"With a tension of, say, fifty Daniell cells, no appreciable quantity will pass across the dust of blacklead or fine charcoal powder loosely arranged, even when the battery poles are approached very near to one another.

"If the tension be increased to, say, two or three hundred cells, the particles arrange themselves by electrical attraction close to one another, making good electrical contact, and forming a channel or bridge through which the electric current freely passes.

"When the tension was still further increased to six or seven hundred cells the author found the electricity would pass from one pole to the other through a considerable interval of the ordinary dust which we get in our rooms, and which is chiefly composed of minute particles of silica and alumina mixed with more or less carbonaceous and earthy matters.

"Incandescent matter offers a very free passage to electrical discharge, as is indicated by the following experiments. The author placed masses of powdered blacklead and powdered wood charcoal in two small crucibles ; no current would pass through these masses whilst they were cold, however close the poles were approached, without actually touching. The battery employed in this experiment was only twelve cells.

"The crucibles were then heated to a red heat, and electricity freely passed through the heated powder; and on testing the resistance opposed by the heated particles, placing the poles 1 inch apart, and employing only six cells, the average resistance opposed by the blacklead was only four British Association units, and that opposed by the wood charcoal five units. The average resistance of a needle telegraph coil may be taken at 300 units, or ohms as they are now termed.

"These observations go to show that an interval of dust separating two metallic conductors opposes practically a decreasing resistance to an increasing electrical tension, and that incandescent particles of carbon oppose about $\frac{1}{60}$th part of the resistance opposed by a needle telegraph coil. Reasoning upon these data, the author was led to construct what he terms a 'lightning-bridge,' which he constructs in the following way:—

"Two thick metal conductors terminating in points are inserted usually in a piece of wood. These points approach one another within about $\frac{1}{18}$th of an inch in a chamber cut in the middle of the wood.

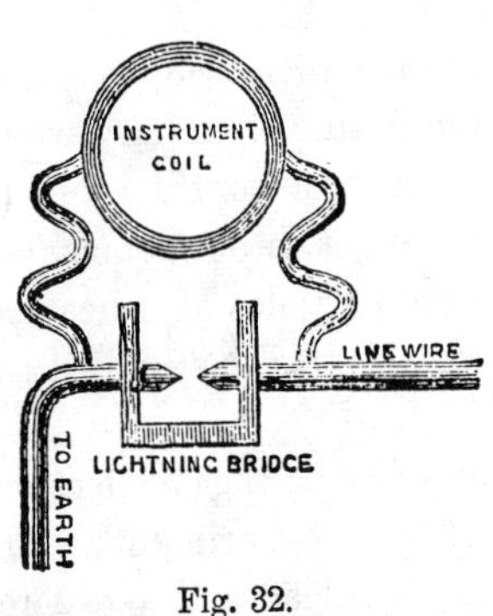

Fig. 32.

"This bridge is placed in the electric circuit in the most direct course which the lightning can take, as shown in the diagram (fig. 32), and the space separating the two points is filled loosely with powder, which is placed in the chamber, and surrounds and covers the extremities of the pointed conductors.

"The powder employed consists of carbon (a conductor) and a non-conducting substance in a minute state of division. The lightning finds in its direct path a bridge of powder, consisting of particles of conducting matter in close proximity to one another; it connects these under the influence of the discharge, and throws the particles into a highly incandescent state. Incandescent matter, as has been already demonstrated, offers a very free passage to electricity, and so the lightning discharge finds an easier passage across the heated matter than through the coils.

"The reason a powder consisting entirely or chiefly of conducting matter cannot be safely employed is that, although in the ordinary conditions of things it would be found to oppose a practically infinite resistance to the passage of electricity of the tension of ordinary working currents, when a high tension discharge occurs the particles under the influence of the discharge will generally be found to arrange themselves so closely as to make a conducting connection between the two points of the lightning-bridge. This can be experimentally demonstrated by allowing the secondary currents developed by a Ruhmkorff's coil to spark through a loose mass of blacklead.[1]

"These lightning-bridges have been in use since January 1866. At the present time there are upwards of one thousand doing duty in this country alone, and not a single case has occurred of a coil being fused when protected by them.

"It is only right, however, to mention that three cases, but three cases only, have occurred where connection was made under the influence of electrical discharge between the two metallic points in the bridge.

"The protectors in which this occurred were amongst those first constructed, in which a larger proportion of conducting matter was employed than the inventor now adopts. The points also in those first constructed were approached to $\frac{1}{50}$th of an inch from one another; and the author has no

[1] See pp. 292, 293 *infra*.

doubt, from an examination of the bridges afterwards, that under the influence of a high tension discharge connection was made between the two metallic points by a bridge of conducting matter, arranged closely together, and if the instruments *had been shaken to loosen the powder*, all would have been put right." [1]

In the little-known researches of the Italian professor, Calzecchi-Onesti, we find this curious phenomenon again

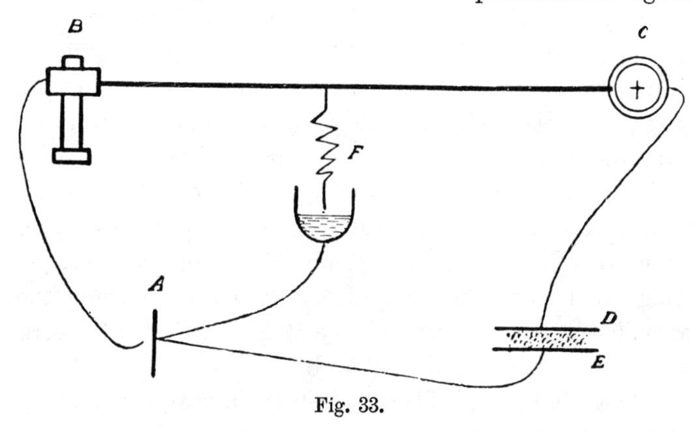

Fig. 33.

cropping up, and in a form more apposite from our present point of view. In 1884-85 Prof. Calzecchi-Onesti found that copper filings heaped between two plates of brass were conductors or non-conductors according to the degree of heaping and pressure, and that in the latter case they could be made conductors under the influence of induction. Fig. 33 illustrates his experiment. In the circuit of a small battery A is placed a telephone B, a galvanometer C, and two brass plates D E, separated by the copper filings. So long as the short-circuit arrangement F (a wire dipping into mercury) is

[1] Sir Wm. Preece tells us the arrangement acted well, but was subject to what we now call coherence, which rendered the cure more troublesome than the disease, and its use had to be abandoned.

open, the galvanometer shows traces of a very feeble current across the filings; but, on dipping the wire for a moment into the mercury and then withdrawing it, a sharp click is heard in the telephone, and the galvanometer indicates the passing of a strong current, showing that the filings must now be conductors. This change he traced to the induced current of the telephone coil (the extra-current *direct*) at the moment of opening the short-circuit. He repeated this experiment with various powders or filings of metal, and ended by showing that rapid interruptions of a circuit containing an inductance coil, or contact with an electrified body, or electro-static discharges were sufficient to make the filings conductive.

For these experiments Calzecchi-Onesti had actually constructed a glass tube (35 millimetres long and 10 millimetres internal diameter) only differing from that shown in fig. 34 in that it was revolvable on its axis, for the purpose of, as we now say, decohering the particles, one revolution or less of the tube sufficing for this purpose.

These observations were published in 'Il Nuovo Cimento,' October 15, 1884, and March 2, 1885,[1] but attracted no attention; and it was only after Prof. E. Branly, of the Catholic University of Paris, had published his results in 1890 that the earlier discoveries of Varley and Onesti came to be remembered and appreciated at their proper value.

Prof. Branly's investigations are very clearly described in 'La Lumière Électrique, May and June 1891.'[2] As this now classic paper deals with facts which are at the very

[1] See also 'Jour. Inst. Elec. Engs.,' vol. xvi. p. 156. In March 1886 Calzecchi-Onesti suggested the use of his tube as a detector of seismical movements, thinking that the conductivity of the filings, imparted by one or other of the above means, would be destroyed by even the smallest earth movement.

[2] See also an abstract in the 'Electrician,' vol. xxvii. pp. 221, 448.

foundation of the Marconi system, I give some extracts from it in Appendix C. Here, therefore, I need only say that Branly verified and extended Calzecchi-Onesti's observations, and made the further (and for our purpose *vital*) discovery that conducting power was imparted to filings by electric discharges in their vicinity, and that this power can be destroyed by simply shaking or tapping them.

The Branly detector, as constructed by Prof. Lodge, is shown in fig. 34. It consists of an ebonite or glass tube about 7 inches long, half-an-inch outer diameter, and fitted at the ends with copper pistons, which can be regulated to press on the filings with any required degree of pressure.

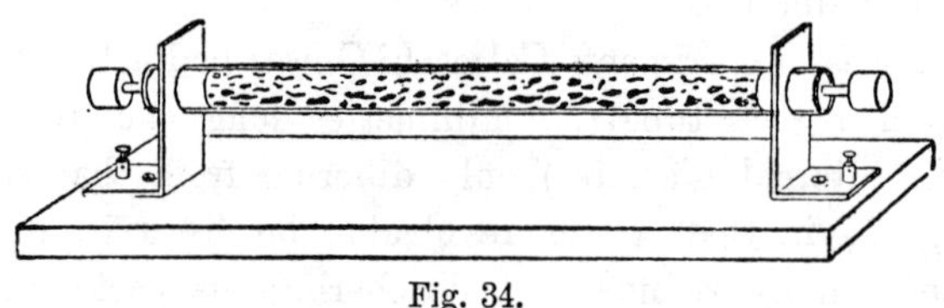

Fig. 34.

To bring back the filings to their normal non-conducting state, Lodge applied to the tube a mechanical tapper, worked either by clockwork or by a trembling electrical mechanism.

These, then, the exciters and the detectors of Hertzian waves, are the bricks and mortar, so to speak, of the Marconi system, and it now only remains to see how they have been shaped and put together to produce a telegraph without connecting wires, which is the realisation of the dream of Steinheil in 1838. And, first, we must notice two or three applications, or suggested applications, which preceded the announcement of Marconi's invention. We do so without in the least meaning to detract one iota from the merit due to the young Irish-Italian inventor,[1] for we believe the idea

[1] Guglielmo Marconi was born in Bologna, 25th April 1874, and was educated at Leghorn, and at the Bologna University, where he was a sedulous attendant at the lectures of Prof. A. Righi.

was entirely original with him, and was unprompted by any suggestions from outside. The history of the applications of science to art shows us that these applications often occur simultaneously to several persons, and it is, therefore, not strange that such is the case in the present instance.

Sir William Crookes, the eminent chemist and electrician, was, I believe, the first to distinctly foresee the applicability of Hertzian waves to practical telegraphy. In a very interesting paper on "Some Possibilities of Electricity,"[1] he gives us the following marvellous forecast of the Marconi system:—

"Rays of light will not pierce through a wall, nor, as we know only too well, through a London fog; but electrical vibrations of a yard or more in wave-length will easily pierce such *media*, which to them will be transparent. Here is revealed the bewildering possibility of telegraphy without wires, posts, cables, or any of our present costly appliances. Granted a few reasonable postulates, the whole thing comes well within the realms of possible fulfilment. At present experimentalists are able to generate electric waves of any desired length, and to keep up a succession of such waves radiating into space in all directions. It is possible, too, with some of these rays, if not with all, to refract them through suitably shaped bodies acting as lenses, and so to direct a sheaf of rays in any given direction. Also an experimentalist at a distance can receive some, if not all, of these rays on a properly constituted instrument, and by concerted signals messages in the Morse code can thus pass from one operator to another.

"What remains to be discovered is—firstly, simpler and more certain means of generating electrical rays of any

[1] 'Fortnightly Review,' February 1892, p. 173. Prof. Lodge has since kindly pointed out to me that about 1890 Prof. R. Threlfall of Sydney, N.S. Wales, threw out a suggestion of the same kind at a meeting of the Australasian Association for the Advancement of Science.

desired wave-length, from the shortest, say a few feet, which will easily pass through buildings and fogs, to those long waves whose lengths are measured by tens, hundreds, and thousands of miles; secondly, more delicate receivers which will respond to wave-lengths between certain defined limits and be silent to all others; and thirdly, means of darting the sheaf of rays in any desired direction, whether by lenses or reflectors, by the help of which the sensitiveness of the receiver (apparently the most difficult of the problems to be solved) would not need to be so delicate as when the rays to be picked up are simply radiating into space, and fading away according to the law of inverse squares. . . .

"At first sight an objection to this plan would be its want of secrecy. Assuming that the correspondents were a mile apart, the transmitter would send out the waves in all directions, and it would therefore be possible for any one living within a mile of the sender to receive the communication. This could be got over in two ways. If the exact position of both sending and receiving instruments were known, the rays could be concentrated with more or less exactness on the receiver. If, however, the sender and receiver were moving about, so that the lens device could not be adopted, the correspondents must attune their instruments to a definite wave-length, say, for example, 50 yards. I assume here that the progress of discovery would give instruments capable of adjustment by turning a screw, or altering the length of a wire, so as to become receptive of waves of any preconcerted length. Thus, when adjusted to 50-yard waves, the transmitter might emit, and the receiver respond to, rays varying between 45 and 55 yards, and be silent to all others. Considering that there would be the whole range of waves to choose from, varying from a few feet to several thousand miles, there would be sufficient secrecy, for the most inveterate curiosity would surely recoil from the task of passing in review all the millions of possible wave-lengths on the remote chance of ultimately hitting on the particular wave-length employed by those whose correspondence it was wished to tap. By coding the message even this remote chance of surreptitious tapping could be rendered useless.

"This is no mere dream of a visionary philosopher. All the requisites needed to bring it within the grasp of daily life are well within the possibilities of discovery, and are so reasonable and so clearly in the path of researches which are now being actively prosecuted in every capital of Europe, that we may any day expect to hear that they have emerged from the realms of speculation into those of sober fact. Even now, indeed, telegraphing without wires is possible within a restricted radius of a few hundred yards, and some years ago I assisted at experiments where messages were transmitted from one part of a house to another without an intervening wire by almost the identical means here described."[1]

In 1893 Nikola Tesla, the lightning-juggler, proposed to transmit electrical oscillations to any distance through space, by erecting at each end a vertical conductor, connected at its lower end to earth and at its upper end to a conducting body of large surface. Owing to press of other work this experiment was never tried, and so has remained a bare suggestion.[2]

At the Royal Institution, June 1, 1894, and later in the

[1] The experiments here referred to were made in 1879 by Prof. Hughes, who has kindly supplied the author with an account of them. As this interesting and important document was received too late for embodiment in the text, I must ask my readers to refer to Appendix D.

[2] See a full account of Tesla's marvellous researches in 'Jour. Inst. Elec. Engs.' for 1892, No. 97, p. 51; also 'Pearson's Magazine,' May 1899, for some of his latest wonders.

same year at the Oxford meeting of the British Association, Prof. Lodge showed how his form of Branly detector could be made to indicate signals at a distance of about 150 yards from the exciter, but at this time the applicability of his experiment to practical long-distance telegraphy was hardly grasped by him. Referring to this in his 'Work of Hertz' (p. 67, 1897 edition), he says:—

"Signalling was easily carried on from a distance through walls and other obstacles, an emitter being outside and a galvanometer and detector inside the room. Distance without obstacle was no difficulty, only free distance is not very easy to get in a town, and stupidly enough no attempt was made to apply any but the feeblest power so as to test how far the disturbance could really be detected.

"Mr Rutherford, however, with a magnetic detector of his own invention, constructed on a totally different principle, and probably much less sensitive than a coherer, did make the attempt (June 1896), and succeeded in signalling across half a mile full of intervening streets and houses at Cambridge."

Between 1895 and 1896 Messrs Popoff, Minchin, Rutherford, and others applied the Hertzian method to the study of atmospheric electricity; and their mode of procedure, in the use of detectors in connection with vertical exploring rods, was much the same as that of Marconi.

Popoff's arrangement especially is so like Marconi's that we are tempted to reproduce it from the 'Elektritchestvo' of St Petersburg for July 1896. Fig. 35 shows the apparatus, the action of which is easily understood. The relay actuates another circuit, not shown, containing a Richard's register, which plots graphically the atmospheric perturbations.

Prof. Popoff's plans were communicated to the Physico-Chemical Society of St Petersburg in April 1895; and in a further note, dated December 1895, he adds: "I entertain the hope that when my apparatus is perfected it will be applicable to the transmission of signals to a distance by means of rapid electric vibrations—when, in fact, a sufficiently powerful generator of these vibrations is discovered." We shall see presently that Popoff was looking in the wrong

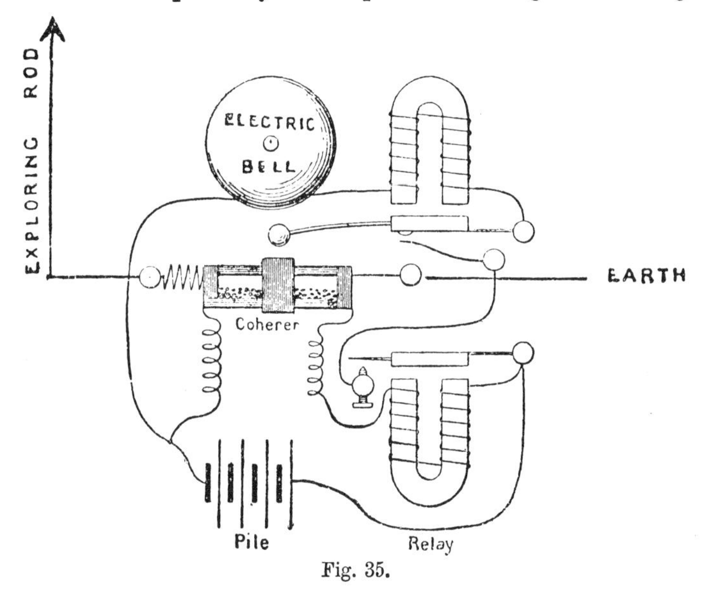

Fig. 35.

direction. It was not so much a more powerful generator (which is easily obtained) that was wanted, as a detector more suitable for signalling purposes than the Branly-Lodge arrangement which he used. Mr Marconi, we shall see, supplied this, and in doing so did the main thing necessary to make Popoff's apparatus a *practical* telegraph.[1]

[1] On hearing of Marconi's success in England, Prof. Popoff tried his apparatus *quasi* telegraph (presumably using more sensitive detectors), and in April 1897 succeeded in signalling through a space of 1 kilometre, then through $1\frac{1}{2}$, and finally through 5 kilometres, with vertical wires, 18 metres high.

Sir Wm. Preece tells us that in December 1895 Captain Jackson, R.N., commenced working in the same direction, and succeeded in getting Morse signals through space before he heard of Marconi. His experiments, however, were treated as confidential at the time, and have not been published.

In 1896 the Rev. F. Jervis-Smith had a detector made of finely-powdered carbon, such as is used in incandescent electric lamps (in fact, a kind of carbon-powder telephone), for observing atmospheric electricity; and a little later (in the spring of 1897) he actually applied it to telegraphic purposes over a distance of more than a mile. This form of detector was to a certain extent self-restoring and did not require any tapping device.[1]

Finally, in 1896, Mr Charles A. Stevenson, of whose work in wireless telegraphy we have already spoken (p. 119, *supra*), had the idea of utilising the coherer principle in the construction of a relay of great delicacy.[2] He does not, however, enter into details, merely referring to his "relay with metallic powder between two electro-magnets" in the course of some remarks on Prof. Blake's experiments in America (p. 121, *supra*).

I now come to Mr Marconi, whose special application of Hertzian waves to practical telegraphy will be easily understood if my readers have carefully followed me in the preceding pages.

His apparatus for short distances, with clear open spaces, consists of the parts which are shown in diagrammatic form

[1] Recently, October 1898, I have seen it stated that Signor Rovelli has found that a detector made of iron filings acts well, and requires no tapping. See also Prof. Chunder Bose's important researches on potassium as a self-restoring detector—'Proceedings Royal Society,' July 1899.

[2] 'Electrical Review,' August 1896.

in figs. 36, 37, 38, and 39. The apparatus at the sending station consists of a modified Righi exciter A (fig. 36), a Ruhmkorff coil B, a battery of a few cells C, and a Morse key K.

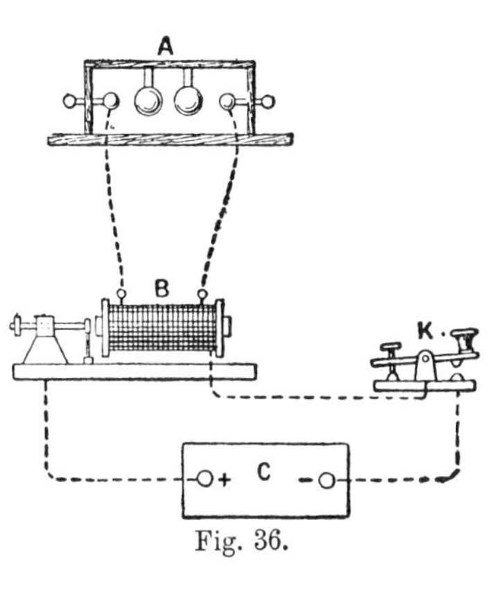

Fig. 36.

The exciter consists of two solid brass spheres A B (fig. 37), 11 centimetres in diameter and 1 millimetre apart. The spheres are fixed in an oil-tight case of parchment or ebonite, so that an outside hemisphere of each is exposed, the other hemispheres being immersed in vaseline-oil thickened by the addition of a little vaseline. As already explained, the use of oil has several advantages, all of which combine to

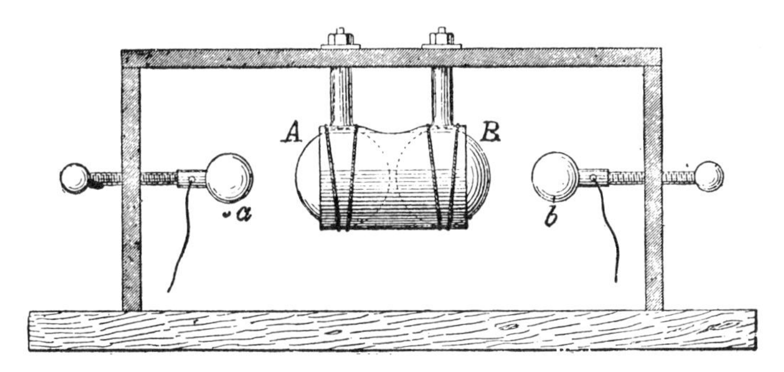

Fig. 37.

increase the effectiveness of the arrangement, and therefore the distance at which the effect can be detected. It keeps the opposing surfaces of the spheres clean and bright, and gives to the electric sparks a more uniform and regular character, which is best adapted for signalling.[1] Two small balls, also of solid brass, *a b*, are fixed in a line with the large ones, usually about 2·5 centimetres apart, and are capable of adjustment. The larger the spheres and balls, and the greater the distances separating them (compatible with the power of the induction coil), the higher is the potential of the sparks and the greater the oscillations to which they give rise, and consequently the greater the distance at which they are perceptible. The balls *a b* are connected each to one end of the secondary coil of the Ruhmkorff apparatus B. The primary wire of the induction coil is excited by the battery C, thrown in and out of circuit by the key K. The efficiency of the sending apparatus depends greatly on the power and constancy of the induction coil: thus a coil yielding a 6-inch spark will be effective up to three or four miles; but for greater distances than this more powerful coils, as one emitting 10-inch sparks, must be used.[2]

The various parts of the sending apparatus are generally so constructed and adjusted as to emit per second about 250 million waves of about 1·3 metres long.

At the receiving station N (fig. 38) is Marconi's special form of the Branly-Lodge detector, shown full size in fig. 39. This is the part which gave him the most trouble. While

[1] Mr Marconi's later experience has led him to doubt these advantages, and to discard the use of oil. He now uses simply a single spark-gap between two balls, as *a b* in fig. 37. See 'Jour. Inst. Elec. Engs.,' No. 139, p. 311, or p. 232 *infra*.

[2] But there is a limit: powerful induction coils of the Ruhmkorff kind are difficult to make and keep in order, and do not by reason of their residual magnetism admit of the very rapid make-and-break action required. Doubtless other and more effective means of excitement will soon be discovered, as Tesla's oscillators, or by the use of Wehnelt's electrolytic contact-breaker, which can be made to interrupt a current one thousand times and more per second. See 'Jour. Inst. Elec. Engs.,' No. 131, p. 317.

for laboratory experiments any detector sufficed to give indications on a sensitive mirror galvanometer at a distance of a few yards, Mr Marconi had to seek a thoroughly practical and reliable arrangement which could stand the comparatively rough usage of everyday work, be restorable to its

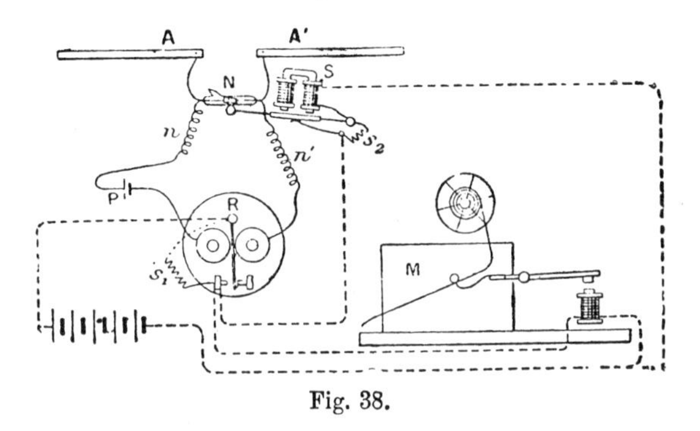

Fig. 38.

normal condition (after every wave) with the utmost certainty, and, at the same time, be sufficiently responsive to the very feeble waves which are found at a great distance from the source, so as to allow of the passage of a current strong enough to actuate a telegraph relay. His detector consists

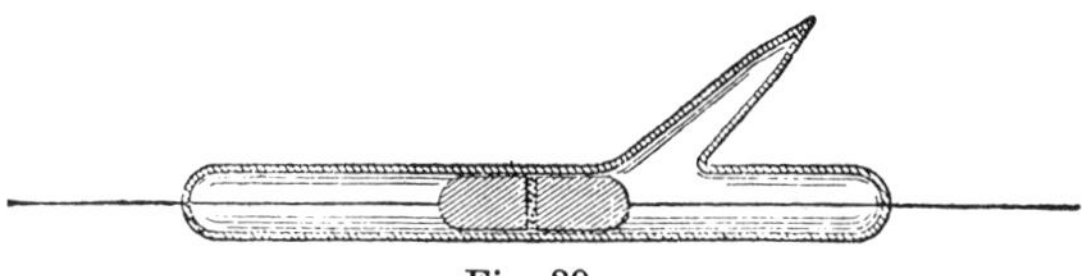

Fig. 39.

of a glass tube, 4 centimetres long and 2·5 millimetres interior diameter, into which two silver pole-pieces, 1 millimetre apart, are tightly fitted, so as to prevent any scattering of the powder. The small intervening space is filled with a mixture of 96 parts of nickel and 4 of silver, not too finely powdered, and worked up with a trace of mercury.

By increasing the proportion of silver powder the sensitiveness of the detector is increased *pro rata;* but it is better for ordinary work not to have too great sensitiveness, as the detector then too readily responds to atmospheric electricity and other stray currents. Similarly, the smaller the powder space the more sensitive is the instrument; but if too small, the action is capricious. The quantity of powder required is, of course, very small, but it must be treated with care: it must neither be too compressed nor too loose. If too tight the action is irregular, and often the particles will not return to their normal condition, or "decohere," as Lodge expresses it; if too loose coherence is slight, and the instrument is not sufficiently sensitive. The best adjustment is obtained when the detector works well under the action of the sparks from a small electric trembler at one metre's distance. The tube is then hermetically sealed, having been previously exhausted of air to about $\frac{1}{1000}$th of an atmosphere. This, though not essential, is desirable, as it prevents the oxidation of the powder.

In its normal condition the metallic powder, as already stated, is practically a non-conductor, offering many megohms resistance. The particles (to use Preece's expressive words) lie higgledy-piggledy, anyhow, in disorder. They lightly touch each other in a chaotic manner; but when electric waves fall upon them they are polarised—order is installed—they are marshalled in serried ranks and press on each other,—in a word, they cohere, electrical continuity is established, and a current passes, the resistance falling from practical insulation to a few ohms or a few hundred ohms according to the energy of the received impacts. Usually it ranges from 100 to 500 ohms.[1]

[1] The action of the detector is hardly yet understood, but recent investigations of Arons (Broca, 'Télégraphie sans Fils,' Paris, 1899, p. 117), of Sundorph ('Science Abstracts,' No. 23, p. 757), and of

The detector is included in the circuit of two electro-magnetic impedance or choking coils *n n'*, a local battery of one or two Leclanché cells P, and a fairly sensitive polarised relay as ordinarily used in telegraphy R. The impedance or choking coils, consisting of a few turns of insulated copper wire on a glass tube, containing an iron bar 5 or 6 centimetres long, are intended to prevent the electric energy escaping through the relay circuit. Prof. Silvanus Thompson doubts the efficacy of this contrivance, but Mr Marconi's experience shows its great utility. Thus, when the coils are removed, all other things remaining the same, the signalling distance is reduced by nearly one-half.

A A' are resonance plates or wings (copper strips) whose dimensions must be adjusted so as to bring the detector into tune electrically with the exciter.

The relay actuates two local circuits on the parallel or shunt system, one containing an ordinary Morse instrument M, and the other the tapper S. The relay and tapper are provided with small shunt coils s_1 and s_2 to prevent sparking at the contacts, which would otherwise impair the good working of the detector. The Morse instrument and the tapper may also be connected in series in one circuit, in which case the former may be made to act as a buzzer, the signals being read by sound. Indeed, the Morse machine may be left out altogether and the signals be read from the sound of the tapper alone. The printing lever of the Morse is so adjusted—an easy matter—as not to follow the rapid makes and breaks of the local current caused by the action of the tapper. Consequently, although the current in the

Tommasina ('Electrician,' vol. xliv. p. 213) seem to bear out the view adopted in the text. Compare Prof. Lodge's views *re* coherence in his 'Work of Hertz,' pp. 22, 70. Also Lamotte's excellent article on "Cohéreurs ou Radioconducteurs," 'L'Éclairage Électrique,' Paris, March 31, 1900.

coils of the Morse is rapidly discontinuous, the lever remains down (and prints) so long as the detector is influenced by the waves sent out by the exciter. In this way the lever gives an exact reproduction of the movements of the distant sending key, dots and dashes at the key coming out as dots and dashes in the Morse. The speed at which signalling can be carried on is but little slower than that in ordinary (Morse) telegraphy, fifteen words a minute being easily attained.

In practice, the sending part of the apparatus should be screened as much as possible by interposed metal plates from the receiving instruments, so as to prevent local inductive interferences; or better, the detector may be shut up in a metal box.

This arrangement is effective for short distances, up to two miles, with clear open spaces, especially if metallic reflectors are erected behind the exciter and detector, and carefully focussed so as to throw the electric rays in the right direction. But for long distances, and where obstacles intervene, as trees, houses, hills — in fact, for practical purposes — certain modifications are necessary which are shown in fig. 40. Reflectors are discarded which are troublesome and expensive to make and difficult to adjust. One knob of the exciter is connected to a stout insulated copper wire, led to the top of a mast and terminating in a square sheet or a cylinder of zinc, which Marconi calls a "capacity area." For still greater distances the wire may be flown from a kite or balloon[1] covered with tinfoil.

[1] In a recent popular lecture it is seriously stated that, when kites are used to carry the conductors, "the electricity obtained from the air, when they were flown high enough, was sufficient to enable the operator to do away with a primary battery" ! ('Electrical Engineer,' October 1, 1897). This is the Mahlon Loomis idea *redivivus* (see p. 68 *supra*), and is as true as another "vulgar error"—to wit, that Marconi, and now Tesla, can explode torpedoes and powder-magazines

The other knob of the exciter is connected to a good earth.

The exciting apparatus is adapted and adjusted for the emission into space of much longer waves than those mentioned on page 208. The wave-length is determined by the height of the vertical wire, being approximately equal to

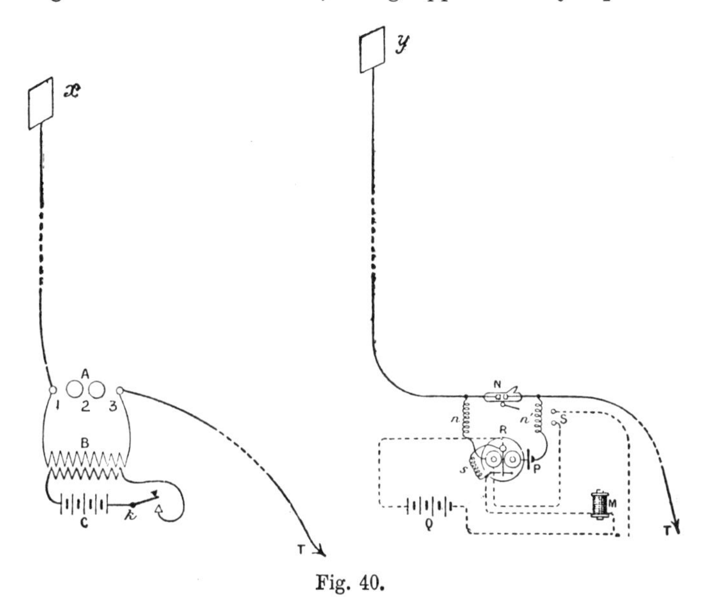

Fig. 40.

four times the height, so that in long-distance signalling the Marconi waves may be many hundreds of feet long.

At the receiving station the resonance wings of the detector are discarded, and one side is connected to a vertical wire and the other side to earth, as in the case of the exciter. Of course, in practice only one vertical wire is required at

at their own sweet will. This, of course, might be done, *if* they could plant a properly adjusted exploding apparatus near the powder; but if they could do this, they could, as Preece says, do many other funny things.

each station, as by means of a switch it can be connected with the exciter for sending, or with the detector for receiving, as may be necessary. The parallelism of the wires and plates, x and y, should be preserved as much as possible in order to obtain the best effects.

The *raison d'être* of the earth connections is not yet clearly understood. An earth wire on the exciter for *long* distances is essential, but at the detector it may apparently be dispensed with without any (appreciable) effect.[1]

However this may be, an earth wire (and a good one too) should be used on the detector as well as on the exciter, if only as a protection from lightning. The vertical wire is practically a lightning-catcher, and the detector is an excellent lightning-guard *when* connected to earth. But if disconnected from earth, and lightning strikes the wire, then we may expect all the disastrous results which follow from a badly constructed or defective lightning-protector. The fear, then, that the Marconi apparatus is especially dangerous may be put aside. Being an excellent lightning-conductor and lightning-guard in one, it may, in my opinion, be safely used, even in a powder-magazine.

From a long series of experiments in Italy in 1895 Mr Marconi worked out a law of distance which all his later experience seems to verify. "The results," he says, "showed that the distance at which signals could be obtained varied approximately as the square of the height of the capacity areas from earth, or, perhaps, as the square of the length of the vertical conductors. This law furnishes us with a safe means of calculating what length the vertical wire should be in order to obtain results at a given distance. The law has never failed to give the expected results *across clear space* in any installation I have carried out, although it usually seems that the distance actually obtained is slightly

[1] 'Jour. Inst. Elec. Engs.,' No. 137, pp. 801, 802, 900, 918, 946, 962.

in excess. I find that, with parity of other conditions, vertical wires 20 feet long are sufficient for communicating one mile, 40 feet four miles, 80 feet sixteen miles, and so on.

"Professor Ascoli has confirmed this law, and demonstrated mathematically, using Neumann's formula, that the action is directly proportional to the square of the length of one of the two conductors if the two are vertical and of equal length,[1] and in simple inverse proportion to the distance between them. Therefore the intensity of the received oscillation does not diminish with the increase of distance if the length of the vertical conductors is increased in proportion, or as the square root of the distance."[2]

Delicate as the apparatus undoubtedly is, and complicated as it may seem, its action is simplicity itself to the telegraphist, differing only in the kind of electricity and the medium of communication from that of the everyday telegraph. On depressing the key *k* (fig. 40) to make, say, a dash, induced currents are set up in the secondary coil of the Ruhmkorff machine; the vertical wire is thereby "charged" up to such a point that it "discharges" itself in sparks across the gaps 1, 2, and 3, and this charging and discharging goes on with extreme rapidity. The wire thus becomes the seat of a rapidly alternating or oscillating current, which gives rise to an equally rapid oscillatory disturbance of the ether all round the wire. These ether oscillations are the Hertzian waves, and

[1] If of unequal lengths then the action is proportional to the product of the two lengths, which, however, must not be too dissimilar. Thus, in the recent American Navy trials, signals from a torpedo-boat with 45 feet of vertical wire to a warship with 140 feet of wire were read at a distance of eighty-five miles; but *vice versâ*, from the higher sending to the lower receiving wire, signalling was only practicable over seven miles. See p. 243 *infra*.

[2] Recent experience goes to show that there is no such simple law. Greater distances are now worked over with shorter wires than formerly.

they spread out into space, much as water waves do when a stone is thrown into a pond, or as air waves do when a sound or a musical note is struck. On arriving at the receiving station these Hertzian, or, as they are also called, electro-magnetic waves, enfeebled more or less as the distance is great or small, strike the wire *y*, and generate along it an oscillatory current of the same kind (though, of course, weaker) as that along the wire *x*. This results in what I may call invisible sparks across the detector gap, which break down the insulation resistance of the contained powder and make it conductive, thus allowing the local battery to act; the relay thereupon closes, and the Morse instrument sounds, or prints the signal as may be required, the tapper all the while doing its work of decohering.

This account of what occurs on depressing the key must be considered as popular rather than as scientifically accurate, for I do not think we yet know what actually takes place, or precisely how it takes place. It must also be confessed that the Marconi apparatus itself is still in the empirical stage, and many questions connected with its distinctive features and their interdependence have yet to be solved. For instance, is the Marconi effect under all circumstances truly Hertzian and oscillatory? Some authorities seem to think that it is one of electro-static, others of electro-magnetic, induction. Again, do the waves radiating from the sending station always travel in rectilinear lines, or are they susceptible of deflection by intervening masses of earth and water? To obtain the best effects, the elevated wires must be vertical as regards the earth, and parallel to each other; but how can they be both in the case of great distances where the curvature of the earth comes into play? Are the capacity areas *x* and *y* necessary? Some say no; others, and amongst them Mr Marconi, say yes, but only for short distances. Then again, assuming that true Hertzian waves are radiated from x and arrive at y, how do the feeble invisible sparks (so to speak) which they evoke at the detector gap act upon the filings so as to make them conductive? Why is it that transmission is practicable to greater distances over sea than over land? Why is a thick vertical wire better for use with the exciter, and a thin wire for use with the detector? Finally, why is it (apparently) immaterial whether or not we use an earth connection on the detector? These are some of the questions awaiting solution; but if I may hazard an opinion, I would say that when solved we shall find that after all the Marconi effect is but on a large scale a Leyden jar effect, complicated no doubt, but still such as every schoolboy is familiar with in principle, and that it conforms to the same laws and conditions.

Marconi's first trials on a small scale were made at Bologna, and these proving successful he came to England and applied for a patent, June 2, 1896.[1] Soon after, in July, he submitted his plans to the postal-telegraph authorities, and, to his honour be it said, they were unhesitatingly —even eagerly—taken up by Preece, although, as we have already seen, he was introducing a method of his own.

The first experiments in England were from a room in the General Post Office, London, to an impromptu station on the roof, over 100 yards distant, with several walls, &c., intervening. Then, a little later, trials were made over Salisbury Plain for a clear open distance of nearly two miles. In these experiments roughly-made copper parabolic reflectors were employed, with resonance plates on each side of the detector (see figs. 36, 38).

[1] This being the first patent of the *New* Telegraphy order, is historically interesting. I have therefore thought it convenient to reproduce it in Appendix E, with the original rough drawings.

In May 1897 still more extensive trials were made across the Bristol Channel between Lavernock and Flat Holm, 3·3 miles, and between Lavernock and Brean Down, near Weston-super-Mare, 8·7 miles (see fig. 20, *supra*). Here the reflectors and resonance plates were discarded. Earth and vertical air wires were employed, as in fig. 40, the vertical wires being in the first case 50 yards high, while in the second case kites carrying the wires were had recourse to.

The receiving apparatus was at first set up on the cliff at Lavernock Point, about 20 yards above sea-level. Here was erected a pole, 30 yards high, on the top of which was a cylindrical cap of zinc, 2 yards long and 1 yard diameter. Connected with this cap was an insulated copper wire leading to one side of the detector, the other side of which was connected to a wire led down the cliff and dipping into the sea. At Flat Holm the sending apparatus was arranged, the Ruhmkorff coil used giving 20-inch sparks with an eight-cell battery.

On the 10th May experiments on Preece's electro-magnetic method (already fully described) were repeated, and with perfect success.

The next few days were eventful ones in the history of Mr Marconi. On the 11th and 12th his experiments were unsatisfactory—worse, they were failures—and the fate of the new system trembled in the balance. An inspiration saved it. On the 13th the receiving apparatus was carried down to the beach at the foot of the cliff, and connected by another 20 yards of wire to the pole above, thus making a height of 50 yards in all. Result, magic! The instruments, which for two days failed to record anything intelligible, now rang out the signals clear and unmistakable, and all by the addition of a few yards of wire! Thus often, as Carlyle says, do mighty events turn on a straw.

Prof. Slaby of Charlottenberg, who assisted at these experiments, has told us in a few graphic words the feelings of those engaged. "It will be for me," he says, "an ineffaceable recollection. Five of us stood round the apparatus in a wooden shed as a shelter from the gale, with eyes and ears directed towards the instruments with an attention which was almost painful, and waited for the hoisting of a flag, which was the signal that all was ready. Instantaneously we heard the first *tic tac*, *tic tac*, and saw the Morse instrument print the signals which came to us silently and invisibly from the island rock, whose contour was scarcely visible to the naked eye—came to us dancing on that unknown and mysterious agent the ether!"

After this the further experiments passed off with scarcely a hitch, and on the following day communication was established between Lavernock and Brean Down.

The next important trials were carried out at Spezia, by request of the Italian Government, between July 10 and 18, 1897. The first three days were taken up with experiments between two land stations 3·6 kilometres apart, which were perfectly successful. On the 14th, the sending apparatus being at the arsenal of San Bartolomeo, the receiving instruments were placed on board a tug vessel, moored at various distances from the shore. The shore wire was 26 metres high, and could be increased to 34 if necessary; the tug wire was carried to the top of the mast, and was 16 metres high. The results were unsatisfactory: signals came, but they were jumbled up with other weird signals, which came from the atmosphere (the weather was stormy) in the way which telegraph and telephone operators know so well. On the 15th and 16th (the weather having moderated) better results were obtained, and communication was kept up at distances up to 7·5 kilometres.

On the 17th and 18th the receiving apparatus was transferred to a warship (ironclad), and, with a shore elevation of

34 metres and a ship elevation of 22 metres, signals were good at all distances up to 12 kilometres, and fairly so at 16 kilometres.

During these experiments it was observed that whenever the funnels, iron masts, and wire ropes of the vessels were in line with the shore apparatus the detector did not work properly, which was to be expected from the screening property of metals; but another and more serious difficulty was also encountered. When the vessel got behind a point of the land which cut off the view of the shore station, the signals came capriciously, and good working was not established until the shore was again in full view. Here was a difficulty which must be surmounted if the new system was to be of any practical utility. We have seen in our account of the work of Hertz that electric waves pass without appreciable hindrance through doors and walls and, generally, non-conducting bodies, being only arrested by metals and other conductors; but in practice, when we come to deal with doors and walls in large masses—as trees, buildings, hills—they seem to partake of the nature of metals, and largely absorb the waves, just as light passes through a thin sheet of glass but is arrested by a thick sheet.

This is one of the vexed questions connected with the theory of the Marconi telegraph. In the early days intervening obstacles certainly did interfere with correct signalling, and in some cases they do so still.[1] Yet in many of Marconi's later trials he appears to have found no difficulty. At the Isle of Wight a hill 300 feet higher than his vertical wires has proved no obstacle.

In the experiments at Dover during the last British Association meeting (August 1899) the great mass of the

[1] 'Jour. Inst. Elec. Engs.,' No. 139, pp. 295, 305, 315; 'Science Abstracts,' No. 15, p. 214, and No. 24, p. 878; 'Electrician,' vol. xliv. pp. 140, 212.

Castle Rock, 400 feet high, did not seem to interfere with the signalling between Dover Town Hall and the South Foreland lighthouse, four miles distant, or the Goodwin lightship, twelve miles farther off. Again, between the Town Hall and Wimereux, across Channel, a mass of houses, tall buildings, and overhead tramway wires appeared to have no bad effect.[1]

Better proof still, we learn that during the same experiments the Wimereux signals intended for Dover were received at the Marconi factory at Chelmsford, eighty-five miles distant from the French station, and that, in fact, signalling was carried on between those two places.[2]

During the naval manœuvres last summer (1899) off Bantry, messages were correctly exchanged between ships when a hill over 800 feet high intervened; and, again, between the Europa and Juno, when eighty-five miles apart, and with thirty ironclads, &c. (with all their masses of metal, funnels, iron masts, and wire rigging), manœuvring in between. The vertical wire on each ship was 170 feet high, so that, owing to the curvature of the earth, a hill of water must have intervened, through or round which the electric waves must have travelled—but which?

According to the observations of Le Bon,[3] they must have gone round it. The length, he says, of the Hertzian waves enables them to turn round obstacles with facility, even metallic bodies in certain circumstances—a fact which accounts for the, apparently, partial transparence of metallic mirrors. "Non-metallic bodies," he goes on to say, "have been considered to be perfectly transparent to Hertzian waves, but do these waves go through a hill or round it? 12 centimetres of Portland cement are only partially

[1] 'Electrician,' vol. xliii. pp. 737, 768.
[2] 'Electrician,' vol. xliii. p. 816.
[3] 'Science Abstracts,' No. 22, p. 671.

transparent, while 30 centimetres are non-transparent or wholly opaque. Dry sand is almost entirely transparent, but wet sand much less so—that is, is partially opaque. Freestone is more transparent than cement, but increases in opacity as it becomes wet. Generally speaking, the transparency of non-metallic bodies varies for each substance and decreases as the thickness and humidity of the body increase." If this be so, the Hertzian waves which act upon a detector on the other side of a hill must go over and round the hill, not through it, just as they go round the edges of metallic mirrors, or travel over the bent or looped wire in some experiments of Hertz and Lodge.

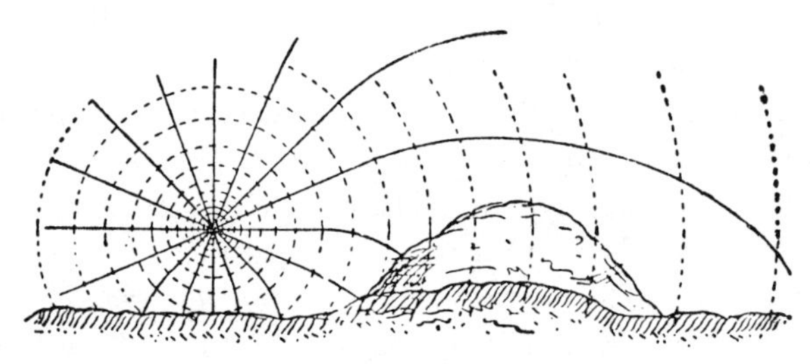

Fig. 41.

This is also the conclusion at which Sir William Preece,[1] Mr Marconi himself,[2] and other authorities have arrived. When, says the former, the ether is entangled in matter of different degrees of inductivity, the lines of force are curved, as in fact they are in light. Fig. 41, which I borrow from Preece, shows how, according to his view, hills are bridged over.

On the other hand, Prof. Branly, while maintaining the theory that electric waves travel in straight lines only, has thrown out the suggestion that the opacity or otherwise

[1] Lecture, Royal Institution, June 4, 1897.

[2] Lecture, Royal Institution, February 2, 1900. Compare his view in 'Jour. Inst. Elec. Engs.,' March 2, 1899.

of intervening bodies may be only a question of wave-lengths—that such bodies may be opaque to some waves, and transparent or partially so to others. Referring to the proposal for firing submarine mines from a distance by means of electric waves, he says, "The thing can only be done if water is transparent to the waves used. The fact that a sheet of tinfoil is capable of completely intercepting electric waves, would make us think that the opacity of water, and especially of salt water, is very probable. He tested various liquids and solutions experimentally, and found that a layer of tap water, 20 centimetres thick, suffices to reduce the signalling distance to one-fifth of its value in open air. The same thickness of salt water intercepts the waves completely. Mineral oil is no more absorptive than air itself. Sea salt is particularly absorptive —more so than the sulphates of zinc, sodium, and copper. The result is therefore fatal to the use of electric waves across intervening water; but it is just possible that the wave-length used may make some difference. Waves from a 2-cm. spark are completely intercepted, while those from a 20-cm. Righi spark are transmitted to the extent of about $\frac{1}{2}$ per cent by sea water 20 cm. thick. It should be remembered that sea water is largely transparent to electric waves of the length of light waves [Röntgen waves], and it is just possible that there are other regions of non-absorption in the electric spectrum." [1]

Whatever the explanation may be, the fact remains that intervening masses do reduce the distance over which a given power and adjustment of apparatus can work, and that their effect is greater over land than over sea—by about one-third. When therefore it is said that interposed bodies offer no difficulty, it should be understood that they offer no difficulty that is not surmountable, and

[1] 'Comptes Rendus,' October 1899, quoted in the 'Electrician,' vol. xliv. p. 140.

we may suppose that the loss is in practice compensated for in one or both of the following ways: (1) by increasing the height of the vertical wires, and so increasing the length of the wave and the volume of the ether disturbed at the sending station; and (2) by increasing the power of the sending and the sensitiveness of the receiving apparatus. But we speedily reach a limit in these directions, so that as far as one can see at present the effective distance of the Marconi system must be small compared with the older methods of telegraphy by wire.

Of course, if ever required, means of automatically repeating the signals could be devised, although there would be great practical difficulties attending the use of the metallic screens which would have to be employed. Another young Italian, Mr Guarini-Foresio, is now working in this direction.[1]

On his return to Germany after witnessing the Marconi trials in England, Prof. Slaby in September 1897 engaged in some very instructive experiments in the vicinity of Potsdam, first between Matrosenstation and the church at Sacrow, 1·6 kilometre, and then between the former place and the castle of Pfaueninsel, 3·1 kilometres. I take the following particulars from the 'Electrical Engineer,' December 3, 1897:—

Prof. Slaby recently, at a technical college in Berlin, gave an interesting report of his experiments on telegraphy without wires, or, as he wants it to be called, "spark telegraphy." He mentioned an experiment made by himself by which he was able to send by means of one wire two different messages simultaneously without interfering with each other. He explained that the continuous current used in ordinary telegraphy is conducted along the middle of the wire, and he

[1] See his *brochure*, 'Transmission de L'Électricité sans Fil,' 2nd edition, p. 29 *et seq.*; or 'Electrical Review,' November 10, 1899.

proved that electric waves on their way through the ether are attracted by wires which come in their way, and that they travel along the outside of those wires without influencing the interior. In making use of these observations he succeeded in sending a wave message along the outside of the wire while another message was proceeding through the centre by the continuous current.

Prof. Slaby says that, in conjunction with Dr Dietz, he made many experiments with "spark telegraphy" before Marconi's inventions became known, but did not achieve any important results.[1]

After his return, however, from England he experimented still further. The Emperor of Germany was present at some of these experiments, and put a number of sailors and the large royal gardens at Potsdam at his disposal. The receiver was erected at the naval station and the transmitter on Peacock Island. The first experiments gave no result, because the coherers used were a great deal too sensitive, and contained, among other things, too much silver, and were affected by the electricity in the atmosphere, and in consequence were constantly affected even when no signals were sent from the sending station.

[1] Referring to these experiments in his book, 'Die Funkentelegraphie,' Berlin, 1897, Prof. Slaby handsomely acknowledges Marconi's merits in the following words: "Like many others, I also had taken up this study, but never got beyond the limits of our High School. Even with the aid of parabolic reflectors and great capacity of apparatus I could not attain any further. Marconi has made a discovery. He worked with means the full importance of which had not been recognised, and which alone explain the secret of his success. I ought to have said this at the commencement of my subject, as latterly, especially in the English technical press, the novelty of Marconi's process was denied. The production of the Hertzian waves, their radiation through space, the sensitiveness of the electric eye, all were known. Very good; but with these means 50 metres were attained, but no more."

Further experiments showed that the results increased in the same measure as the sensitiveness of the coherer decreased. Prof. Slaby uses now very rough and jagged nickel filings which have been carefully cleaned and dried. As the receiving station could not be seen from the island, the sending station was removed to a church a little farther away, and the exciter was put between the columns of the portico, while the mast which carried the wire was erected on the spire. The experiments then went very well.

When the sending apparatus was put back a little farther into the church, and the wire was put for about a length of 2 yards parallel with the stone slabs of the floor and a yard and a half above it, it ceased to work properly, because the waves seek the earth. Hence one must not bring the wire too near to the earth, or lay it parallel when near the earth. When the sending apparatus was moved back to the island, it was found that trees near the wire proved an obstacle because they received the waves. Therefore the Professor says that it is best to so arrange that the wires on the receiver and on the transmitter can be seen from each other. Even the sail of a little boat or the smoke from a steamer cause small interruptions, which make the signals more or less indistinct. The waves get through impedimenta, and even through buildings, but there is always much loss. In order to make the wire which was placed on the island more visible from the mainland, it was lengthened from 25 to 65 yards, and placed upon a boat on the river. That did not remedy matters; but when the wire on the receiver was also lengthened to 65 yards very good results followed, showing that the length of the wire is of great importance.

Prof. Slaby next proceeded, early in October, to experiment over an open stretch of country, free from all intervening obstacles, between Rangsdorf (sending station) and

Schöneberg (receiving station), a distance of 21 kilometres. Captive balloons raised to a height of 300 metres were employed. On the first two days the results were disappointing, and the fault was found to be in the vertical conductors, which consisted of the wire cables holding the balloons. With a double telephone wire there was a slight improvement; and eventually, on the 7th October, "fine insulated copper wire of ·46 millimetres diameter was substituted with excellent results."

Correspondence was now always good, except when disturbed by atmospheric discharges (the weather being stormy). At such times the signals were distorted and confused, and often the discharges were so strong as to unpleasantly shock the operators, making it necessary to handle the apparatus with the greatest care.[1] Here is another serious difficulty with which Mr Marconi has to contend, and from which we see no escape short of total suspension of operations during stormy weather—namely, the great liability to accident and derangement, not merely from lightning flashes, to which all telegraph systems are subject, but from all those other electrical disturbances of the atmosphere which have hitherto been of little account. The greater the distance worked over, the higher must be the conductors, and, consequently, the greater must be the danger.

The apparatus used by Prof. Slaby differed somewhat from Marconi's, the following being the more important points:—

1. A Weston galvanometer relay, which, it is curious to note, is our old friend in modern guise, the Wilkins' relay, used by Mr Wilkins in his wireless telegraph experiments in 1845 (see p. 39, *supra*).

[1] See also Brett's remarks, 'Jour. Inst. Elec. Engs.,' No. 137, p 945; and Preece, 'Jour. Soc. Arts,' vol. xlvii. p. 522.

2. An ordinary Branly-Lodge detector with hard nickel powder only.
3. No impedance or "choking" coils.[1]

The further course of Marconi's experiments is so succinctly given by the chairman of the Wireless Telegraph Company in a recent address, October 7, 1898, that we cannot do better than follow him.[2]

"A year ago," he says, "when this company was started (July 1897), Mr Marconi happened to be in Italy making experiments for the Italian Government, and for the King and Queen at the Quirinal. On his return to this country, the first long-distance trial was made between Bath and Salisbury. The receiver in this case was given to a post-office official, who went to Bath and by himself rigged up a station, at which he received signals thirty-four miles distant from where they were sent at Salisbury. After this we put a permanent station at Alum Bay, Isle of Wight. This station at first was used in connection with a small steamer that cruised about in the neighbourhood of Bournemouth, Boscombe, Poole Bay, and Swanage, a distance of eighteen miles from the Needles Hotel station, with which it was in constant telegraphic communication.

"Various exhibitions were given later—one at the House of Commons, where a station was erected, and another station at St Thomas's Hospital opposite (May 1898). Within an hour of the time our assistants arrived to put up the installation, the system was at work. We had many exhibitions at our offices, at which a number of people

[1] About this time Dr Tuma of Vienna was engaged on similar experiments, using, however, instead of a Ruhmkorff coil a Tesla oscillator or exciter, with nickel powder only in the detector. I have not seen any detailed account of these experiments.

[2] I have incorporated a few passages from Mr Marconi's recent paper (Institution of Electrical Engineers, March 2, 1899), so as to make the account more complete. These are shown in brackets thus [ ].

attended; amongst others Mr Brinton, a director of the Donald Currie line of steamers, who asked if we could report a ship passing our station. This was done. The ship was the Carisbrooke Castle, on her first voyage out, and as she passed the Needles a message reporting the fact was wirelessly telegraphed to Bournemouth, and there put on the ordinary telegraph wires for transmission to Mr Brinton.

"After this Lord Kelvin visited our station at Alum Bay, and expressed himself highly pleased with all he saw. He sent several telegrams, *viâ* Bournemouth, to his friends, for each of which he insisted on paying one shilling royalty, wishing in this way to show his appreciation of the system and to illustrate its fitness for commercial uses. The following day the Italian Ambassador visited the station. Among other messages, he sent a long telegram addressed to the Aide-de-camp to the King of Italy. As it was in Italian, and as Mr Marconi's assistant at Bournemouth had no knowledge of that language, it may be taken as a severe test—as, in fact, a code message. The telegram was received exactly as it was sent. Previously, we had a display for the 'Electrical Review' and the 'Times,' both of which papers sent representatives. They put the system to every possible test, and, among others, sent a long code message, which had to be repeated back. In their reports they stated that this was done exactly as sent.

[In May Lloyd's desired to have an illustration of the possibility of signalling between Ballycastle and Rathlin Island in the north of Ireland. The distance between the two positions is seven and a half miles, of which about four are overland and the remainder across the sea, a high cliff also intervening between the two positions. At Ballycastle a pole 70 feet high was used to support the wire, and at Rathlin a vertical conductor was supported by the lighthouse 80 feet high. Signalling was found quite possible between the two points, but it was thought desirable to bring the height of the pole at Ballycastle to 100 feet, as the proximity of the lighthouse to the wire at Rathlin seemed to diminish the effectiveness of that station. At Rathlin we found that the lighthouse-keepers were not long in learning how to work the instruments, and after the sad accident which happened to poor Mr Glanville, that installation was worked by them alone, there being no expert on the island at the time.[1]]

"Following this, in July last (1898) we were requested by a Dublin paper, the 'Daily Express,' to report the Kingstown regatta. In order to do this we erected a [land] station at Kingstown, and another on board a steamer which followed the yachts. A telephone wire connected the Kingstown station with the 'Daily Express' offices, and as the messages came from the ship they were telephoned to Dublin and published in successive editions of the evening papers.[2]

[After the races longer distances were tried, and it was found that with a height of 80 feet on the ship and 110 feet on land it was possible to communicate up to a distance of twenty-five miles; and it is worthy of note in this case that the curvature of the earth intervened very considerably at such a distance between the two positions.]

"After this, Mr Marconi was requested to put up a station at Osborne to connect with the Prince of Wales' yacht Osborne. Bulletins of the Prince's health (his Royal Highness, as we all know, met with a lamentable accident just

[1] Mr Glanville, a promising young electrician (only twenty-five years old), was missing from Saturday to the Tuesday evening following, when his body, terribly mutilated, was found at the foot of a cliff 300 feet high in Rathlin Island.

[2] Very full illustrated accounts of this remarkable experiment are given in the Dublin 'Mail,' July 20, 21, and 22, 1898.

before then) were reported to her Majesty: not only that, but the royalties made great use of our system during the Cowes week.

[In this installation induction-coils capable of giving a 10-inch spark were used at both stations. The height of the pole supporting the vertical conductor was 100 feet at Osborne House. On the yacht the top of the conductor was attached to the mainmast at a height of 83 feet from the deck, thus being very near one of the funnels, and in the proximity of a great number of wire stays. The vertical conductor consisted of a $\frac{7}{20}$ stranded wire at each station. The yacht was usually moored in Cowes Bay at a distance of nearly two miles from Osborne House, the two positions not being in sight of each other, the hills behind East Cowes intervening.

[On August 12 the Osborne steamed to the Needles and communication was kept up with Osborne House until off Newton Bay, a distance of seven miles, the two positions being completely screened from each other by the hills lying between. From the same position we found it quite possible to speak with our station at Alum Bay, although Headon Hill, Golden Hill, and over five miles of land lay directly between. Headon Hill was 45 feet higher than the top of our wire at Alum Bay, and 314 feet higher than the wire on the yacht.]

"Within the last few days we have had to move our station at Bournemouth four miles farther west, where we have put up the same instruments, the same pole, and everything at the Haven Hotel, Poole, which is eighteen miles from Alum Bay. This increase of distance has no detrimental effect on our work; in fact it seems rather easier, if anything, to receive signals at the Haven Hotel than at our former station: thus, the height of the conductor at Bournemouth was 150 feet, but this is now reduced to 100 feet, which is a very great improvement.[1]

[The vertical conductors are stranded $\frac{7}{20}$ copper wire insulated with india-rubber and tape. A 10-inch spark induction coil is used at each station, worked by a battery of 100 Obach cells M size, the current taken by the coil being 14 volts of from 6 to 9 ampères. The sparks take place between two small spheres about 1 inch diameter, this form of transmitter having been found more simple and more effective than the Righi exciter previously used. The length of spark is adjusted to about 1 centimetre, which, being much shorter than the coil can give, allows a large margin for any irregularity that may occur. No care is now taken to polish the spheres at the place where the sparks occur, as working seems better with dull spheres than with polished ones.]

"The Marconi invention is the only (electric) telegraph by means of which a moving object can be kept in communication with any other moving object, or a fixed station, and therefore any one can see the great use of the invention, not only to the Royal Naval authorities, but also to the mercantile marine. A ship fitted with Mr Marconi's apparatus can not only keep in telegraphic communication with the shore up to any reasonable distance—it has been thoroughly tested up to twenty-five miles off the shore — but ships can also, if properly equipped, be warned of approaching danger or their proximity to dangerous coasts which are fitted with the wireless apparatus.

[If we imagine a lighthouse provided with a transmitter constantly giving an intermittent series of electric waves, and a ship provided with a receiving apparatus placed in the focal line of a reflector, it is plain that when the

[1] The height has since been gradually reduced to 75 feet.

receiver comes within the range of the transmitter the bell will be rung only when the reflector is directed towards the transmitter. If, then, the reflector is caused to revolve by clockwork or by hand, it will give warning only when occupying a certain sector of the circle in which it revolves. It is therefore easy for a ship in a fog to make out the exact direction of the lighthouse, and, by the conventional number of taps or rings corresponding to the waves emitted, she will be able to discern, either a dangerous point to be avoided, or the port for which she is endeavouring to steer.[1]]

[In December of last year the Company thought it desirable to demonstrate that the system was available for telegraphic communication between lightships and the shore. This, as you are aware, is a matter of much importance, as all other systems tried so far have failed, and the cables by which ships are connected are exceedingly expensive, and require special moorings and fittings, which are troublesome to maintain and liable to break in storms. The officials of Trinity House offered us the opportunity of demonstrating to them the utility of the system between the South Fore-

[1] Theoretically this is possible, but practically I fear the size and management of the reflector would make it very difficult. A simpler way *might* be by reverting to the original form of the apparatus (p. 206 *supra*), and by revolving a cylindrical metallic screen (with a longitudinal slit or opening not too wide) around the detector until the position is found in which the bell rings under the influence of the electric rays entering at the opening. Even here I foresee difficulties. However, the thing is easily put to actual test, and, considering its great importance, I am surprised that this has not been done.

Bela Schäfer in Austria, and Russo d'Asar in Italy, are said to be able to determine the presence and course of a ship at 60 to 80 kilometres distant. If this has been done, then, *vice versâ*, a ship should be able to determine the presence and direction of a lighthouse.

land Lighthouse and one of the following light-vessels—viz., the Gull, the South Goodwin, and the East Goodwin. We naturally chose the one farthest away—the East Goodwin—which is just twelve miles from the South Foreland Lighthouse.

[The apparatus was taken on board in an open boat and rigged up in one afternoon. The installation started working from the very first, December 24, without the slightest difficulty. The system has continued to work admirably through all the storms, which during this year have been remarkable for their continuance and severity. On one occasion, during a big gale in January last, a very heavy sea struck the ship, carrying part of her bulwarks away. The report of this mishap was promptly telegraphed to the superintendent of Trinity House, with all details of the damage sustained.

[The height of the wire on board the ship is 80 feet, the mast being for 60 feet of its length of iron, and the remainder of wood. The aërial wire is led down among a great number of metal stays and chains, which do not appear to have any detrimental effect on the strength of the signals. The instruments are placed in the aft-cabin, and the aërial wire comes through the framework of a skylight, from which it is insulated by means of a rubber pipe. As usual, a 10-inch coil is used, worked by a battery of dry cells, the current taken being about 6 to 8 ampères at 14 volts.

[The instruments at the South Foreland Lighthouse are similar to those used on the ship; but as we contemplate making some long-distance tests from the South Foreland to the coast of France, the height of the pole is much greater than would be necessary for the lightship installation alone.]

These tests were duly carried out, and on March 27,

1899, communication was successfully established between England and France.[1]

"On this side of the Channel," says the 'Daily Graphic' (March 30, 1899), "the operations took place, by permission of the Trinity House, in a little room in the front part of the engine-house from which the power is derived for the South Foreland lighthouses. The house is on the top of the cliffs overlooking the Channel. The demonstrations are being conducted for the benefit of the French Government, who have the system under observation, and besides Signor Marconi there were present at the Foreland yesterday Colonel Comte du Bontavice de Heussey, French Military Attaché in England; Captain Ferrie, representing the French Government; and Captain Fieron, French Naval Attaché in England. During the afternoon a great number of messages in French and English crossed and recrossed between the little room at the South Foreland and the Châlet D'Artois, at Wimereux, near Boulogne.

"The whole of the apparatus stood upon a small table about 3 feet square, in the centre of the room. Underneath the table the space was fitted with about fifty primary cells; a 10-inch induction coil occupied the centre of the table. The spark is 1½ centimetre long, or about three-quarters of an inch; the pole off the top of which the current went into space is 150 feet high. The length of spark and power of current were the same as used for communication with the East Goodwin lightship, a fact which seems remarkable when it is considered that the distance over which the messages were sent yesterday was nearly three times as great. The greater distance is compensated for by the increased height of the pole.

"Throughout the whole of the messages sent yesterday

[1] All the London daily papers of March 29 and 30 contain full and glowing accounts of this installation.

there was not once a fault to be detected—everything was clearly and easily recorded. The rate of transmission was about fifteen words a minute."

The first international press message sent by the new system was secured by the 'Times,' and is as follows:—

"(*From our Boulogne Correspondent.*)

"WIMEREUX, *March* 28.

"Communication between England and the Continent was set up yesterday morning by the Marconi system of wireless telegraphy. The points between which the experiments are being conducted are South Foreland and Wimereux, a village on the French coast two miles north of Boulogne, where a vertical standard wire, 150 feet high, has been set up. The distance is thirty-two miles. The experiments are being carried on in the Morse code. Signor Marconi is here conducting the trials, and is very well satisfied with the results obtained.

"This message has been transmitted by the Marconi system from Wimereux to the Foreland."

Amongst the experts in electrical science who witnessed these experiments was Prof. Fleming, F.R.S., of University College, London, who has given us his impressions in a long letter to the 'Times' (April 3, 1899). He tells us that throughout the period of his visit messages, signals, congratulations, and jokes were freely exchanged between the operators sitting on either side of the Channel, and automatically printed down in telegraphic code signals on the ordinary paper slip at the rate of twelve to eighteen words a minute. Not once was there the slightest difficulty or delay in obtaining an instant reply to a signal sent. No familiarity with the subject removes the feeling of vague wonder with which one sees a telegraphic instrument merely connected with a length of 150 feet of copper wire run up the side of a flagstaff begin to draw its message out of space and print down in dot and dash on the paper tape the intelligence ferried across thirty miles of water by the mysterious ether.

An extensive trial of the system between ships at sea was next made during the British naval manœuvres in July 1899. Three ships of the B fleet were fitted up —the flagship, Alexandra, and the cruisers, Juno and Europa. The greatest distance to which signals were sent

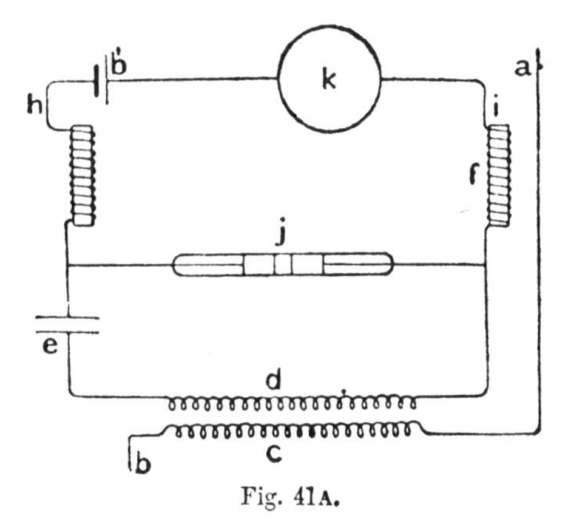

Fig. 41A.

was sixty nautical miles between the Juno and Europa, and forty nautical miles between the Juno and Alexandra. These were not the maximum ranges attained, but the distances at which, under all circumstances, the system could be relied on for certain and accurate transmission. Test signals were obtainable up to a distance of seventy-four nautical miles (eighty-five miles).

These important results were obtained by the use of Marconi's peculiar form of induction coil or transformer.[1]

[1] Then just patented. See his specification, No. 12,326, of June 1, 1898 (accepted July 1, 1899); or abstract in 'Electrician,' vol. xliii. p. 847.

Fig. 41A shows the arrangement, where *a* is the vertical wire, *b* the earth connection, *c* the primary and *d* the secondary wires of the transformer, and *e* a condenser. "The object of this arrangement," says Mr Marconi, "is to increase the electromotive force of the oscillations at the terminals of the detector *j*, and therefore to cause its state of insulation to break down with weaker oscillations, and so be affected at a much greater distance than is possible when the detector is connected directly with the vertical wire." The primary coil is wound with fine wire (contrary to the usual practice), and the secondary with still finer wire.

In his first experiments with transformer coils of various

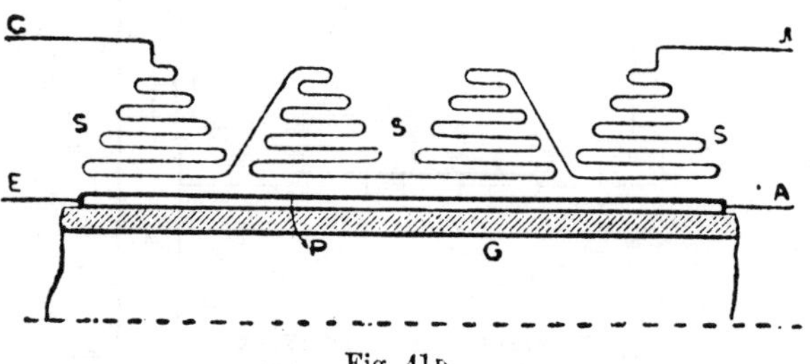

Fig. 41B.

kinds, Marconi found that if the secondary wire be wound in more than one layer, little if any advantage is obtained. He was led to try a mode of winding in which the centre of the coil still consisted of a single layer, but with the number of turns increased at the ends. This gave much better results, and led, finally, to the arrangement shown in fig. 41B, which "represents an enlarged half-longitudinal section of the coil, but is not drawn strictly to scale. Also, instead of showing the section of each layer of wire as a longitudinal row of dots or small circles, as it should appear, it is for simplicity drawn as a single continuous line." It will thus be seen that the secondary wire is wound curiously in four sections; and another peculiarity is that these sections must be connected together in the way shown, and, as the distance from the primary wire increases, the number of turns in each section must decrease.

The use of these small coils during the naval manœuvres had a very marked effect on the detector, enabling it to respond to waves from greater distances. Thus, when working between the Juno and Europa with a given power in the transmitter and height of vertical wires, the effective signalling distance was seven nautical miles without the coil, and sixty nautical miles with it.[1]

After the naval manœuvres Marconi stations were opened at Chelmsford and Harwich, forty miles apart; and in August (1899), during the meetings of the British Association at Dover and the French Association at Boulogne, messages were freely exchanged between the two places, the distance across Channel being about thirty miles. Correspondence was also kept up between Dover and the South Foreland (four miles) and the Goodwin (sixteen miles) stations across the great masses of the Castle Rock (400 feet high) and the South Foreland cliffs. Communication was also found possible between Wimereux and Chelmsford or Harwich. The distance in each of these cases is about eighty-five miles, of which thirty are over sea and fifty-five over land. The height of the vertical wires at each end was 150 feet, thus showing, and confirming the results of many previous experiments, that considerable masses of intervening rock, earth, and water do not offer an insurmountable obstacle to the transmission of signals. If they did, and if it had been necessary for a line drawn between the tops of the wires to clear the curvature of the earth,

[1] 'Electrician,' vol. xliv. p. 555. Prof. Fessenden, in America, uses in the same way a specially constructed transformer which is reported to be many times more effective than Marconi's ('Electrician,' vol. xliii. p. 807), and with which "it should be possible to signal across the Atlantic with 200 feet vertical wires"! ('Globe,' January 1, 1900.)

they would have had to be in this case over 1000 feet high.

In America, October 1899, the Marconi apparatus was employed to report from sea the progress of the yachts in the international contest between the Columbia and the Shamrock. The working was (of course) perfectly satisfactory, and as many as 4000 words are said to have been transmitted in one day from the (two) ship stations to the shore station.[2]

Immediately after the races the instruments were placed, by request, at the service of the American Navy Board, who put them to some severe and interesting tests. The cruiser, New York, and the battleship, Massachusetts, were equipped under Mr Marconi's personal supervision. The two vessels lay at anchor in the North River, 480 yards apart, or about the distance that would separate ships steaming in squadron formation. The signalling operations on the New York were performed by Mr Marconi himself, aided by an assistant, and under the directions of two members of the Navy Board; while the signalling on the Massachusetts was done by one of Marconi's assistants, under the inspection of another navy official. The object of the first experiments was to determine the practicability of the system for short-distance signalling between squadrons at sea. The first test was the sending and receiving of a newspaper article of about 1500 words, which was done without error, and at a speed of eleven words per minute. The second test was the transmission of a series of numbers of various lengths, which was also done correctly, and with a little more rapidity. The third test dealt with a series of letters written down at random; the fourth, a series of short messages; and the fifth and sixth, series of code-word

[1] 'Electrician,' vol. xliii. pp. 737, 768, 793, 816; vol. xliv. p. 557.
[2] 'New York Herald' (Paris edition), October 6, 1899.

messages. These latter naturally taxed the skill of the operators, the "words" having a weird look, unpronounceable, and with absolutely no sense or meaning. It is therefore not surprising that in these tests one or two errors were detected; but they were probably as much the fault of the operator as of the apparatus. Indeed, as Mr Marconi has pointed out, all these experiments were more tests of the operators for correctness and speed of signalling than of the utility of the apparatus, which for such short distances was incontestable.

The vessels then left for the open sea. At a point about five miles off the Highlands the New York anchored, while the Massachusetts continued on her course, exchanging signals with her consort at intervals of ten minutes. Up to some distance short of thirty-six miles the signals were good, but what that distance was the report from which we are quoting does not specify; it merely says, "At a distance of thirty-six miles the messages failed to carry, and the battleship came back and anchored a few hundred yards from the New York."[1]

In order to test the possibility of interference with the signals, a Marconi apparatus was established at the Highlands, with a vertical wire of 150 feet. At intervals during the time that messages were being exchanged between the two warships the Highlands station sent out other signals, with the invariable result that the correspondence between the ships was rendered unintelligible.[2]

The official report of such an independent authority as the American Navy Board must always be valuable; and as, moreover, it contains precise information on other points

[1] In an article in the 'Times,' November 16, 1899, the effective distance is said to have been thirty-five miles, and that the apparatus was only designed to carry thirty miles, that being considered the outside range requisite for the yacht-reporting operations.

[2] 'Electrician,' vol. xliv. p. 106.

not referred to in the preceding paragraphs, I think it useful to reproduce it as follows:—

"We respectfully submit the following findings as the result of our investigation of the Marconi system of wireless telegraphy: It is well adapted for use in squadron signalling under conditions of rain, fog, and darkness. Wind, rain, fog, and other conditions of weather do not affect the transmission; but dampness may reduce the range, rapidity, and accuracy by impairing the insulation of the aerial wire and the instruments. Darkness has no effect. We have no data as to the effects of rolling and pitching; but excessive vibration at high speed apparently produced no bad effect on the instruments, and we believe the working of the system would be very little affected by the motion of the ship. The accuracy is good within the working ranges. Cipher and important signals may be repeated back to the sending station, if necessary, to ensure absolute accuracy. When ships are close together (less than 400 yards) adjustments, easily made, of the instruments are necessary. The greatest distance that messages were exchanged with the station at Navesink was 16½ miles. This distance was exceeded considerably during the yacht races, when a more efficient set of instruments was installed there.[1] The best location of instruments would be below, well protected, in easy communication with the commanding officer. The spark of the sending coil, or of a considerable leak, due to faulty insulation of the sending wire, would be sufficient to ignite an inflammable mixture of gas or other easily lighted matter, but with direct lead (through air space, if possible) and the high insulation necessary for good work no danger of fire need be apprehended. When two

[1] This is a mistake. The instruments were the same in both cases. See 'Times' article, November 16, 1899.

transmitters are sending at the same time, all the receiving wires within range receive the impulses, and the tapes, although unreadable, show unmistakably that such double sending is taking place. In every case, under a great number of varied conditions, the interference was complete. Mr Marconi, although he stated to the Board before these attempts were made that he could prevent interference, never explained how, nor made any attempt to demonstrate that it could be done. Between large ships (heights of masts 130 feet and 140 feet) and a torpedo-boat (height of mast 45 feet), across open water, signals can be read up to seven miles on the torpedo-boat and eighty-five miles on the ship. Communication might be interrupted altogether when tall buildings of iron framing intervene. The rapidity is not greater than twelve words per minute for skilled operators. The shock from the sending coil of wire may be quite severe, and even dangerous to a person with a weak heart. No fatal accidents have been recorded. The liability to accident from lightning has not been ascertained. The sending apparatus and wire would injuriously affect the compass if placed near it. The exact distance is not known, and should be determined by experiment. The system is adapted for use on all vessels of the navy, including torpedo-boats and small vessels, as patrols, scouts, and despatch boats, but it is impracticable in a small boat. For landing-parties the only feasible method of use would be to erect a pole on shore and thence communicate with the ship. The system could be adapted to the telegraphic determination of differences of longitude in surveying. The Board respectfully recommends that the system be given a trial in the navy."[1]

On Mr Marconi's return voyage from America he gave an interesting demonstration of the value of his system for

[1] 'Electrician,' vol. xliv. p. 212.

ships at sea. "A few days previous," he says, "to my departure, the war in South Africa broke out. Some of the officials of the American liner suggested that, as a permanent installation existed at the Needles, Isle of Wight, it would be a great thing, if possible, to obtain the latest war news before our arrival at Southampton. I readily consented to fit up my instruments on the St Paul, and succeeded in calling up the Needles station at a distance of sixty-six nautical miles, when all the important news was received on board, the ship the while steaming her twenty knots per hour. The news was collected and printed in a small paper, called the 'Transatlantic Times,' several hours before our arrival at Southampton."[1]

In October 1899 the War Office sent out some Marconi instruments to South Africa, for use at the base and on the railways; but the military authorities on the spot realised that the system could only be of value at the front, and the apparatus was moved up to the camp at De Aar. The results at first were not altogether satisfactory, a fact which is accounted for by the absence of suitable poles or kites; and afterwards, when kites were improvised, the wind was so variable that it often happened that when the kite was flying at one station there was a calm at the other station. However, when suitable kites were obtained, and the wind was favourable, communication was possible from De Aar to the Orange River, or about seventy miles. Stations were subsequently established at Belmont, Enslin, and Modder river on the west, and in Natal on the east. No reliable reports of the work of these installations

[1] 'Electrician,' vol. xliv. p. 557. Also the 'Times,' November 16 and 18, 1899. This unique production was printed by the ship's compositor, and published at a dollar per copy, the proceeds going to the Seamen's Fund. The 'Times' of November 16 reproduces the contents.

amongst the South African kopjes have yet reached us, but we hope, with Mr Marconi, that "before the campaign is ended wireless telegraphy will have proved its utility in actual warfare." [1]

Having now brought my account of the more important of Marconi's public demonstrations up to date,[2] I propose to occupy a few final pages with some further remarks on the theory and practice of Hertzian-wave telegraphy.

It has been objected to the Marconi system that, with the removal of the reflectors and the resonance wings, the condition of privacy in telegrams is no longer possible, since any one provided with the necessary apparatus can receive the signals at any point within the circle of which the sending station is the centre and the receiving station the radius. Another, and in some cases more serious, objection is that any one by erecting a wire or wires in the vicinity of a Marconi station can propagate therefrom Hertzian waves, which by interference will so confuse the effects in the detector as to make correct signalling impracticable. It may not even be necessary to propagate counter-waves: a large sheet of metal (or several such sheets) erected high in air, in line with the stations, at right angles to the direction of the waves, and connected by a wire to the earth, will intercept much of the energy, and the more so as it is near to either of the stations. Thus, if used for naval or

[1] 'Electrician,' vol. xliv. p. 557. Up to date (January 1901) we have no authentic accounts of the results obtained, although the Marconi staff have long ago returned to England. Rumour says the kopjes have proved too difficult. See, however, Major Flood Page's remarks, 'Electrician,' March 2, 1900.

[2] Of course I do not pretend that these are the only demonstrations of value that have been made. In America, France, Germany, and Italy, and doubtless in other countries, important experiments have been and are being made; but beyond occasional brief notices of them in the newspapers, and still fewer notices in the technical journals, few clear and veracious accounts have come under my notice.

military purposes, an enemy could either tap the dispatches or render them unintelligible at pleasure. The latter objection is from the nature of things unavoidable, and in practice must limit the application of the system to lines of communication sufficiently apart as not to interfere with one another. The first objection, however, can be obviated to some extent by reverting to the condition of syntony or resonance with reflectors, and it is in this direction that improvements may soon be expected.

Dr Oliver Lodge, F.R.S., the distinguished Professor of Physics, University College, Liverpool, and the coadjutor and expounder of Hertz in England, has long been engaged on the problem of a Hertzian-wave telegraph—especially with a view of securing syntony in the sending and receiving apparatus, and thereby limiting the communications to similarly attuned instruments, the absence of which selective character is at present one of the great drawbacks of the Marconi system.

We have seen (p. 204, *supra*) that as early as June 1, 1894, Prof. Lodge had exhibited apparatus which was effective for signalling on a small scale, but, as he says, "stupidly enough no attempt was then made to apply any but the feeblest power, so as to test how far the disturbance could really be detected. . . . There remained, no doubt, a number of points of detail, and considerable improvements in construction, if the method was ever to become practically useful."[1] These he has since worked out, and some of them are embodied in his patent, No. 11,575, of May 10, 1897, "Improvements in Syntonised Telegraphy without Line Wires."

As capacity areas, spheres or square plates of metal may be employed; but for the purpose of combining low resistance with large electro-static capacity, cones or triangles are

[1] 'The Work of Hertz,' pp. 67, 68.

preferred, with the vertices adjoining and their larger areas spreading out into space. Or a single insulated surface may be used in conjunction with the earth—the earth, or conductors embedded in it, constituting the other capacity area. As radiation from these surfaces is greater in the equatorial than in the axial direction, so, when signalling in all directions is desired, the axis of the emitter should be vertical. Moreover, radiation in a horizontal plane is less likely to

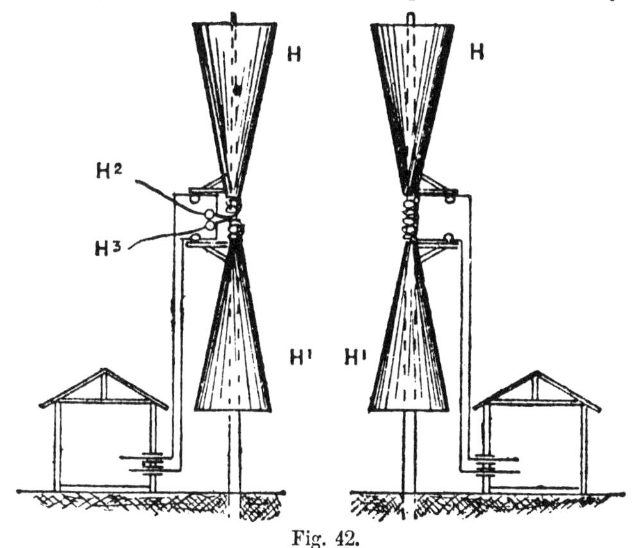

Fig. 42.

be absorbed during its passage over partially conducting earth or water.

Fig. 42 shows the arrangement for long-distance signalling. H H¹ are large triangular sheets of metal, which by means of suitable switches (not shown) can be connected to the sending or the receiving apparatus as desired. Those on the left-hand side of the figure are shown in connection with polished knobs H² H³ (protected by glass from ultra-violet light), which form the adjustable spark-gap of the exciter. Between each capacity area and its knob is inserted a self-inductance coil of thick wire or metallic ribbon (see H^4, fig. 43) suitably insulated, the object of which is to prolong the electrical oscillations in a succession of waves, and thereby obtain a definite frequency or pitch, rendering syntony possible, since exactitude of working depends on the fact that with the emission of a number of successive waves the feeble impulse at the receiving station is gradually strengthened till it causes a perceptible effect, on the well-known principle of sympathetic resonance.

The capacity areas and inductance coils are exactly alike at the two communicating stations, so as to have the same frequency of electrical vibration. This frequency can be altered either by varying the capacity of the Leyden jars used in the exciting circuit, or by varying the number and position of the inductance coils, or by varying both in the proper degree, thus permitting only those stations whose rate of oscillation is the same to correspond.

To actuate the exciter a Ruhmkorff coil may be used, or a Tesla coil, a Wimshurst machine, or any other high tension apparatus.

Fig. 43 shows the details of the arrangement for exciting and detecting the electric waves. When used as a transmitter the receiving circuit is disconnected from the capacity areas by a suitable switch (not shown). Let us first consider the arrangement as a transmitter. Putting the Ruhmkorff coil A in action, it charges the Leyden jars J J, whose outer coatings are connected, first, through a self-inductance coil H^5 of fairly thin wire, so as to permit of thorough charging of the jars; and, second, to the "supply gaps" H^6 H^7. When the jars are fully charged to sparking-point, sparks occur at the "starting-gap" H^8. These precipitate sparks at the "supply gaps," which evoke electrical charges in the capacity areas H H^1. These charges surge through the inductance coils H^4, and spark into each other across the "discharge gap" between the knobs H 2 H 3. This last discharge, according to Prof. Lodge, is the chief agent in starting the oscillations which are the cause of the emitted waves; but it is permissible to close the "discharge gap," and so leave the oscillations to be started by the sparks at the "supply gaps" only, whose knobs must then be polished and protected from ultra-violet light, "so as to supply the electric charge in as sudden a manner as possible."

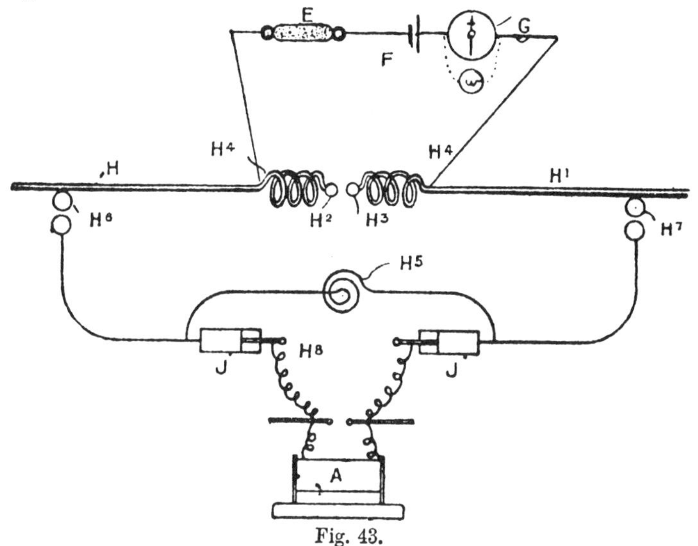

Fig. 43.

When used as a receiver the "discharge-gap" is bridged over by a suitable cut-out, and connection is made with the receiving circuit, as shown on the top of fig. 43. As detector, Lodge uses—

1. His own original form of coherer, fig. 44, wherein a metallic point N rests lightly on a flat metallic surface O (for instance, a needle point of steel or platinum making light contact with a steel or aluminium bar like a watch spring), fixed at one end P, and delicately adjustable by a

micrometer screw Q, so as to regulate the pressure at the point N. Or—

2. A Branly tube filled with selected iron filings of uniform size, sealed up in a good vacuum, and with the

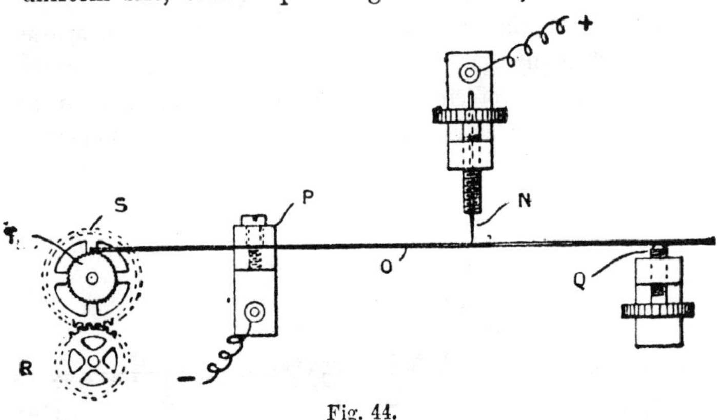

Fig. 44.

electrodes, which are of platinum, reduced to points a short distance apart.

His latest form of the Branly coherer is shown full size in fig. 45, and is said to be exceedingly sensitive and certain in its action, especially in a very high vacuum. A A is a glass tube held tightly by ebonite supports B B; C is a pocket or reservoir for spare filings, which can be added to, or taken from, the effective portion as required by inverting the tube; D D are the *silver* electrodes immersed in the filings, which are, as before, of carefully selected iron of uniform size as nearly as possible; E is one of the terminals of the silver electrodes, the other of which is hidden from view.

The instrument is secured by the clamp screw F to any convenient support, to which the tapping or decohering apparatus is applied.[1]

[1] It appears that to Professor Blondel is due the credit of first constructing a coherer of this kind in August 1898. See the 'Electrician,' vol. xliii. p. 277.

When an electric wave from a distant exciter arrives and stimulates electric vibrations in the syntonised capacity areas, the electrical resistance of the coherer suddenly and greatly falls and permits the small battery F, fig. 43, to actuate a relay G, or a telephone, or other telegraphic instrument.

To break contact, or to restore the original great resistance of the coherer, any form of mechanical vibration suffices, as a clock, or a tuning-fork, or a cog-wheel (as in fig. 44), or other device for causing a shake or tremor, and kept in motion by a spring, or weight, or by electrical means. Indeed, the mere motion of any clockwork attached to the coherer stand will suffice, an exceedingly slight, almost imperceptible, tremor being all that is usually required.

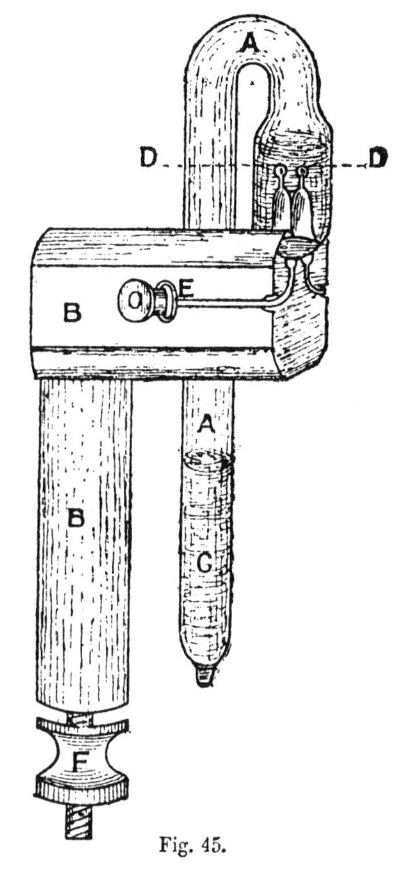

Fig. 45.

Usually the coherer is arranged in simple series with the battery and telegraphic instrument, and is so joined to the capacity areas as to include in its circuit the self-inductance coils—an arrangement which Prof. Lodge considers of great advantage, or, as he says, "an improvement on any mode of connection that had previously been possible without these coils."

The patent specification figures and describes another way—viz., enclosing the inductance coils in an outer or secondary coil (constituting a species of transformer), and making this coil part of the coherer circuit. In this case the coherer is stimulated by the waves in the secondary coil instead of, as before, by those in the inductance coils, which with their capacity areas are thus left free to vibrate without disturbance from attached wires.

In all cases it is permissible, and sometimes desirable, to shunt the coils of the telegraphic instrument G by means of a fine wire or other non-inductive resistance coil *w*, "in order to connect the coherer more effectively and closely to the capacity areas."

At the Royal Society Conversazione on May 11, 1898, a complete set of Lodge's apparatus was shown in action, in which certain modifications in the signalling and recording parts were introduced at the suggestion of Dr Alexander Muirhead. Instead of the ordinary Morse key, Muirhead's well-known automatic transmitter with punched tape was employed at one end of the suite of rooms, and a siphon-recorder as the receiving instrument at the other end. The recorder was so arranged as to print, not as usually zigzag traces, but (the needle working between stops) a momentary deflection mark for a dot and a longer continued mark for a dash.

The siphon-recorder is so quick in its responses that it indicates each one of the group of sparks emitted from the sending apparatus: hence a dash is not merely a deflection held over, but is made up of a series of minute vibrations; and even a dot is seen to consist of similar vibrations, though of course of a lesser number. If the speed of signalling is slow and the recorder tape moves slowly, these vibrations appear as actual dots and dashes; but each signal, when examined with a microscope, is seen to consist of a short or long series of lines representing the constituent vibrations.

At a slow rate of working the signals can thus be got with exceeding clearness; but for actual signalling this is not at all necessary, and it is possible to attain a high speed, making such brief contacts that a single deflection of the recorder needle indicates a dot, and three consecutive deflections a dash. The paper thus marked does not look like the ordinary record, but more resembles the original Morse characters as depicted on pp. 404 and 409 of Shaffner's 'Telegraph Manual' (New York, 1859), and is easily legible with a little practice.

An ordinary telephone was also available as a receiver (connected through a transformer coil) in which the dots and dashes were heard very clearly and distinctly.

The apparatus is reported to have worked well (except at the high speeds, when it occasionally missed fire), and did not seem to be in the least affected by any of the numerous electrical exhibits in the neighbourhood, although some of them must have set up considerable radiation of Hertzian waves.

Based on the same principles—viz., the emission of electric waves at one place and their detection by some form of coherer at another place—there is naturally a similarity in the outlines of the Lodge system and that of Marconi for short distances (where vertical wires are not used), as depicted in fig. 38, *supra*. The differences are differences of arrangement and detail only, but they appear to be fraught with some important consequences.

In the first place, Prof. Lodge claims that his arrangement of the sending apparatus is a more persistent exciter, in that it emits a longer train of longer waves,[1] which by acting cumulatively on the detector breaks down its insulation, when more powerful but fewer trains of shorter waves might be inoperative. Then in the next place, this

[1] For some important observations on this point see Mr A. Campbell Swinton, 'Jour. Inst. Elec. Engs.,' No. 139, p. 317.

element of persistency permits of the use of syntonising contrivances, by means of which the rate of oscillation of any desired set of instruments can be accurately attuned so that only those instruments can correspond, without affecting or being affected by other sets tuned to a different frequency, thus securing to some extent the advantage of privacy in the communications.

Lodge's arrangement has worked well in the laboratory and lecture-room, but he does not appear to have tried it (which is a pity) over any considerable distance, so that it remains to be seen how far he can go without having recourse to vertical wires, which Marconi finds so essential for practical work over distances of more than two or three miles.[1]

Speaking of the waste of energy all round a Marconi transmitter as now constructed, and of the desirability of preventing it if possible, a writer in a recent volume of the 'Electrician' (vol. xli. p. 83) has some remarks which may appropriately be given here. "Unless," he says, "some means are adopted for converging the radiation along a definite path, the practical and commercial efficiency of Hertzian-wave telegraphy will be small, and the enormous quantities of wasted radiation spreading away from the line of signalling will have to be prevented from interfering with other receiving stations. Prof. Lodge has proposed the syntonising of instruments as a means of preventing this interference, and it is undoubtedly possible to tune the receiver so that it will respond only to waves of a particular pitch; but should wireless telegraphy by Hertzian waves ever become extensively practised over considerable distances, the number of possible non-interfering tones of wave-lengths will be found insufficient for the number of receiving stations. Besides, the syntonising method of confining the message to its proper path has the disadvan-

[1] For Professor Lodge's newest developments see his paper, 'Jour. Inst. Elec. Engs.,' No. 137, p. 799, which deserves careful study.

tage that it does not confine the energy to that path; it is therefore very wasteful.

"Hertzian waves, like their natural relatives light waves, have the property that they can be reflected and refracted; though, from the fact of their much greater wave-length, the apparatus requisite for converging them in a parallel beam is more difficult to construct and more costly than is, for example, the parabolic reflector of a search light or the compound lens of a lighthouse. Nevertheless there are well-known substances, of which pitch is an example, which, when formed into a lens or prism, have the power of acting upon Hertzian waves precisely as lenses or prisms of glass act upon rays of ordinary light. As a scientific fact this has been known since Hertz's time, but there would appear to be considerable difficulty in its application.

"We are inclined to think, however, that it will ultimately be found necessary to employ, in wireless telegraphy, some such means as a huge pitch lens would afford for collecting the scattering rays from the Hertzian wave generator or oscillator, and refracting them into a beam of almost, if not quite, parallel rays; thus improving, both in efficiency and in penetrative power, this interesting method of propagating signals through space."

Mr Marconi has been steadily working at these problems of syntony and reflection. The latter is, I fear, only possible for short distances, up to a few miles, and with apparatus as originally constructed (p. 206, *ante*). For greater distances necessitating considerable lengths of vertical wire such huge reflectors would be required, and their adjustment would be so difficult as to make the plan practically impossible.

From syntonising methods some promising results have been obtained. In a recent letter to the 'Times' (October 4, 1900) Prof. Fleming has some startling revelations. "For the last two years," he says, "Mr Marconi has not ceased to grapple with the problem of isolating the lines of communication, and success has now rewarded his skill and industry. Technical details must be left to be described by him later on, but meanwhile I may say that he has modified his receiving and transmitting appliances so that they will only respond to each other when properly tuned to sympathy.

"These experiments have been conducted between two stations 30 miles apart—one near Poole in Dorset and the other near St Catherine's in the Isle of Wight. At the present moment there are established at these places Mr Marconi's latest appliances, so adjusted that each receiver at one station responds only to its corresponding transmitter at the other. During a three days' visit to Poole, Mr Marconi invited me to apply any test I pleased to satisfy myself of the complete independence of the circuits, and the following are two out of many such tests: Two operators at St Catherine's were instructed to send simultaneously two different wireless messages to Poole, and without delay or mistake the two were correctly recorded and printed down at the same time in Morse signals on the tapes of the two corresponding receivers at Poole.

"In this first demonstration each receiver was connected to its own independent aerial wire hung from the same mast. But greater wonders followed. Mr Marconi placed the receivers at Poole one on the top of the other, and connected them both to one and the same wire, about 40 feet in length, attached to a mast. I then asked to have two messages sent at the same moment by the operators at St Catherine's, one in English and the other in French. Without failure each receiver at Poole rolled out its paper tape, the message in English perfect on one and that in French on the other. When it is realised that these visible dots and dashes are the results of trains of intermingled electric waves rushing with the speed of light across the intervening 30 miles, caught on one and the same short aerial wire, and disentangled and sorted out automatically by the two machines into intelligible messages in different languages, the wonder of it all cannot but strike the mind.

"Your space is too valuable to be encroached upon by further details, or else I might mention some marvellous results, exhibited by Mr Marconi during the same demonstrations, of messages received from a transmitter 30 miles away and recorded by an instrument in a closed room merely by the aid of a zinc cylinder, 4 feet high, placed on a chair. More surprising is it to learn that, whilst these experiments have been proceeding between Poole and St Catherine's, others have been taking place for the Admiralty between Portsmouth and Portland, these lines of communication intersecting each other; yet so perfect is the independence that nothing done on one circuit now affects the other, unless desired. A corollary of these latest improvements is that the necessity for very high masts is abolished. Mr Marconi now has established perfect independent wireless telegraphic communication between Poole and St Catherine's, a distance of 30 miles, by means of a pair of metal cylinders elevated 25 feet or 30 feet above the ground at each place."

If these latest improvements yield only one-half of the results indicated by Prof. Fleming, the value of Marconi's system will be enormously enhanced and its sphere of utility correspondingly extended.[1] We therefore await with impatience the promised disclosures as to how all these wonderful things can be done.

Even should the improvements turn out to be of no great practical value, or to be not susceptible of extensive applica-

[1] In which case we shall have, in future editions, to withdraw or at least to modify some of our remarks as to its present limitations.

tion, we can well be content with the system as described in these pages. It has proved to be practical up to sixty or seventy miles, and within this limit there ought to be a wide and useful field for activity. Thus, many outlying islands are within this distance from each other and from the continents, with which communication at *all* times has hitherto been practicable only by the use of cables, which are always costly to make and lay, and often costly to keep in repair. Here, especially between places where the traffic is not great, is a large field to be occupied as cables grow old and fail.

Then, we have seen from the address of the chairman of the Wireless Telegraph Company that negotiations are going on with Lloyd's which, if carried into practical effect, will result in an extensive application for signalling between Lloyd's stations and outward and inward bound vessels passing in their vicinity. Indeed it is not rash to predict that the lighthouses and lightships around the coasts, not only of the British Isles but of all countries, will in time be supplied with wireless telegraphs, keeping up constant correspondence with all who go down to the sea in ships. Then, again, there is the application to intercommunication between ships at sea. Ships carrying the Marconi apparatus can carry on a definite conversation with the occupants of lighthouses and lightships and with each other. It will readily be seen that this might, in many cases, be far more serviceable than the few light signals now obtainable, or the signalling by flags, horns, &c.—a tedious process at best, and one that is often full of uncertainty, if not of positive error.[1]

[1] The English, American, German, and French naval authorities are now making independent experiments with the Marconi system, and it is probable we may soon hear of its adoption, or of some modification of it, as part of the equipment of not only warships but of all large vessels.

Turning from sea to land, we find, for the reasons we have already indicated, a more circumscribed field of application—at all events, until means are devised for focussing the electric rays and rendering the apparatus syntonic. But even then, although by these means we will be able to record messages only where intended, there still remain cross interferences of which I fear we can never be rid, and therefore we can never use the system in a network of lines as now, where wires cross, recross, and overlap each other in all ways and directions. The various waves of electricity would so interfere with each other in their effects on the detectors that the result would be chaos. Therefore wireless telegraphy can only be used in lines removed from each other's disturbing influences, as in sparsely populated countries and undeveloped regions.

However, many cases of impromptu means of communication arise where, as Prof. Lodge says, it might be advantageous to "shout" the message, spreading it broadcast to receivers in all directions, and for which the wireless system is well adapted, seeing that it is so inexpensive and so easily and rapidly installed,—such as for army manœuvres, for reporting races and other sporting events, and, generally, for all important matters occurring beyond the range of the permanent lines.

But for the regular daily correspondence of a nation with its lines ramifying in all directions and carrying enormous traffics, the Marconi system is not adapted, no more than any other wireless method that has been proposed, or is likely to be invented in our day. So, for a long time to come we must keep to our present telegraphic and telephonic wires, using the wireless telegraph as an adjunct for special cases and contingencies such as I have mentioned.

A few words as to the future, by way of conclusion, and

our task is completed. On this point we find some recent remarks of Prof. Silvanus Thompson so appropriate that we quote them in full, as being more authoritative than anything we could ourselves say. Prof. Thompson has thoroughly studied the subject, and therefore "speaks by the card."

"It has been shown," he says, "that there are three general methods of transmitting electric signals across space. All of them require base lines or base areas. The first—conduction—requires moist earth or water as a medium, and is for distances under three miles the most effective of the three. The second—induction—is not dependent upon earth or water, but will equally well cross air or dry rock. The third—electric wave propagation—requires no medium beyond that of the ether of space, but is interfered with by interposed things such as masts or trees. Given proper base lines or base areas, given adequate methods of throwing electric energy into the transmitting system, and sufficiently sensitive instruments to pick up and translate the signals, it is possible, in my opinion, so to develop each of the three methods that by any one of them it will be possible to establish electric communication between England and America across the intervening space. It is certainly possible, either by conduction or by induction; whether by waves I am somewhat less certain. Conduction might very seriously interfere with other electric agencies, since the waste currents in the neighbourhood of the primary base line would be very great. It is certainly possible either by conduction or induction to establish direct communication across space with either the Cape, or India, or Australia (under the same assumptions as before), and at a far less cost than that of a connecting submarine cable.

"Instruments which operate by means of alternating currents of high frequency, like Mr Langdon-Davies's phonophore, are peculiarly liable to set up disturbance in other circuits. A single phonophore circuit can be heard in lines a hundred miles away. When this first came to my notice it impressed me greatly, and coupled in my mind with the Ferranti incident mentioned above" (see note, p. 144, *supra*), "caused me to offer to one of my financial friends in the City, some eight years ago, to undertake seriously to establish telegraphic communication with the Cape, provided £10,000 were forthcoming to establish the necessary basal circuits in the two countries, and the instruments for creating the currents. My offer was deemed too visionary for acceptance. The thing, however, is quite feasible. The one necessary thing is the adequate base line or area. All the rest is detail."[1]

One word more. A press telegram of April 12, 1899, says: "The Wireless Telegraph Company have been approached by the representative of a proposed syndicate which desires to acquire the sole rights of establishing wireless telegraphic communication between England and America. The directors of the Company will consider the matter at their first meeting, which is fixed for an early date."[2]

Thus I end my task as I began it, with a dream—the self-same dream! As to its realisation in the distant future who can say nay?

> "There are more things in heaven and earth, Horatio,
> Than are dreamt of in our philosophy."

---

[1] 'Journal, Society of Arts,' April 1, 1898.

[2] The syndicate must hurry up, as Mr Nikola Tesla is now on their track with a wireless telegraph that will "stagger humanity." We read ('Electrician,' January 19, 1900) that he is convinced he will soon be able to communicate, not only with Paris, but with every city in the world, and that at a speed of from 1500 to 2000 words per minute! See also p. 239, *supra*, for Prof. Fessenden's great hopes.

APPENDIX A.

THE RELATION BETWEEN ELECTRICITY AND LIGHT —BEFORE AND AFTER HERTZ.

Before Hertz.

SUBSTANCE of a lecture by Prof. Oliver Lodge, London Institution, December 16, 1880.[1]

Ever since the subject on which I have to speak to-night was arranged, I have been astonished at my own audacity in proposing to deal, in the course of sixty minutes, with a subject so gigantic and so profound that a course of sixty lectures would be inadequate for its thorough and exhaustive treatment. I must, therefore, confine myself to some few of the most salient points in the relation between electricity and light, and I must economise time by plunging at once into the middle of the matter without further preliminary.

What *is* electricity? We do not know. We cannot assert that it is a form of matter; neither can we deny it. On the other hand, we cannot certainly assert that it is a form of energy; and I should be disposed to deny it. It may be that electricity is an entity *per se*, just as matter is an entity *per se*. Nevertheless, I can tell you what I mean by electricity by appealing to its known behaviour.

Here is a voltaic battery. I want you to regard it, and all electrical machines and batteries, as kinds of electricity-pumps, which drive the electricity along through the wire very much as a water-pump can drive water along pipes.

[1] Based on a report in 'Design and Work,' February 5, 1881.

While this is going on, the wire manifests a whole series of properties, which are called the properties of the current.

[Here were shown an ignited platinum wire, the electric arc between two carbons, an electric machine spark, an induction coil spark, and a vacuum tube glow. Also a large nail was magnetised by being wrapped in the current, and two helices were suspended and seen to direct and attract each other.]

To make a magnet, then, we only need a current of electricity flowing round and round in a whirl. A vortex or whirlpool of electricity is in fact a magnet, and *vice versâ*. And these whirls have the power of directing and attracting other previously existing whirls according to certain laws, called the laws of magnetism. And, moreover, they have the power of exciting fresh whirls in neighbouring conductors, and of repelling them according to the laws of diamagnetism. The theory of the actions is known, though the nature of the whirls, as of the simple streams of electricity, is at present unknown.

[Here was shown a large electro-magnet and an induction-coil vacuum discharge spinning round and round when placed in its field.]

So much for what happens when electricity is made to travel along conductors—*i.e.*, when it travels along like a stream of water in a pipe, or spins round and round like a whirlpool.

But there is another set of phenomena, usually regarded as distinct and of another order, but which are not so distinct as they appear, which manifest themselves when you join the pump to a piece of glass or any non-conductor and try to force the electricity through that. You succeed in driving some through, but the flow is no longer like that of water in an open pipe; it is as if the pipe were completely obstructed by a number of elastic partitions or diaphragms. The water cannot move without straining and bending these diaphragms, and if you allow it, these strained partitions will recover themselves and drive the water back again. [Here was explained the process of charging a Leyden jar.] The essential thing to remember is that we may have electrical energy in two forms, the static and the kinetic;

and it is therefore also possible to have the rapid alternation from one of these forms to the other, called vibration.

Now we will pass to the second question: What do you mean by light? And the first and obvious answer is, Everybody knows. And everybody that is not blind does know to a certain extent. We have a special sense-organ for appreciating light, whereas we have none for electricity. Nevertheless, we must admit that we really know very little about the intimate nature of light—very little more than about electricity. But we do know this, that light is a form of energy; and, moreover, that it is energy rapidly alternating between the static and the kinetic forms—that it is, in fact, a special kind of energy of vibration. We are absolutely certain that light is a periodic disturbance in some medium, periodic both in space and time—that is to say, the same appearances regularly recur at certain equal intervals of distance at the same time, and also present themselves at equal intervals of time at the same place; that, in fact, it belongs to the class of motions called by mathematicians undulatory or wave motions.

Now how much connection between electricity and light have we perceived in this glance into their natures? Not much truly. It amounts to about this: That on the one hand electrical energy may exist in either of two forms—the static form, when insulators are electrically strained by having had electricity driven partially through them (as in the Leyden jar), which strain is a form of energy, because of the tendency to discharge and do work; and the kinetic form, where electricity is moving bodily along through conductors, or whirling round and round inside them, which motion of electricity is a form of energy, because the conductors and whirls can attract or repel each other and thereby do work.

On the other hand, light is the rapid alternation of energy from one of these forms to the other—the static form where the medium is strained, to the kinetic form when it moves. It is just conceivable then that the static form of the energy of light is *electro*-static—that is, that the medium is *electrically* strained—and that the kinetic form of the energy of light is *electro*-kinetic—that is, that the motion is not ordinary motion, but electrical motion—in fact, that light is an electrical vibration, not a material one.

On November 5 last year there died at Cambridge a man in the full vigour of his faculties—such faculties as do not appear many times in a century—whose chief work had been the establishment of this very fact, the discovery of the link connecting light and electricity, and the proof—for I believe that it amounts to a proof—that they are different manifestations of one and the same class of phenomena,—that light is, in fact, an electro-magnetic disturbance. The premature death of James Clerk-Maxwell is a loss to science which appears at present utterly irreparable, for he was engaged in researches that no other man can hope as yet adequately to grasp and follow out; but fortunately it did not occur till he had published his book on 'Electricity and Magnetism,' one of those immortal productions which exalt one's idea of the mind of man, and which has been mentioned by competent critics in the same breath as the 'Principia' itself.

The main proof of the electro-magnetic theory of light is this: The rate at which light travels has been measured many times, and is pretty well known. The rate at which an electro-magnetic wave disturbance would travel, if such could be generated (and Mr Fitzgerald, of Dublin, thinks he has proved that it cannot be generated directly by any known electrical means), can be also determined by calculation from electrical measurements. The two velocities agree exactly.

The first glimpse of this splendid generalisation was caught in 1845, five-and-thirty years ago, by that prince of pure experimentalists, Michael Faraday. His reasons for suspecting some connection between electricity and light are not clear to us—in fact, they could not have been clear to him; but he seems to have felt a conviction that if he only tried long enough, and sent all kinds of rays of light in all possible directions across electric and magnetic fields in all sorts of media, he must ultimately hit upon something. Well, this is very nearly what he did. With a sublime patience and perseverance which remind one of the way Kepler hunted down guess after guess in a different field of research, Faraday combined electricity, or magnetism, and light in all manner of ways, and at last he was rewarded with a result—and a most out-of-the-way result it seemed. First, you have to get a most powerful magnet, and very strongly excite it; then you have to pierce its two poles with holes, in order that a beam of light may travel from one to the other along the lines of force; then, as ordinary light is no good, you must get a beam of plane polarised light and send it between the poles. But still no result is obtained until, finally, you interpose a piece of a rare and out-of-the-way material which Faraday had himself discovered and made, a kind of glass which contains borate of lead, and which is very heavy or dense, and which must be perfectly annealed.

And now, when all these arrangements are completed, what is seen is simply this, that if an analyser is arranged to stop the light and make the field quite dark before the magnet is excited, then directly the battery is connected and the magnet called into action a faint and barely perceptible brightening of the field occurs, which will disappear if the analyser be slightly rotated. [The experiment was shown.] Now, no wonder that no one understood this result. Faraday himself did not understand it at all. He seems to have thought that the magnetic lines of force were rendered luminous, or that the light was magnetised; in fact he was in a fog, and had no idea of its real significance. Nor had any one. Continental philosophers experienced some difficulty and several failures before they were able to repeat the experiment. It was, in fact, discovered too soon, and before the scientific world was ready to receive it, and it was reserved for Sir William Thomson briefly, but very clearly, to point out, and for Clerk-Maxwell more fully to develop, its most important consequences.

This is the fundamental experiment on which Clerk-Maxwell's theory of light is based; but of late years many fresh facts and relations between electricity and light have been discovered, and at the present time they are tumbling in in great numbers.

It was found by Faraday that many other transparent media besides heavy glass would show the phenomenon if placed between the poles, only in a less degree; and the very important observation that air itself exhibits the same phenomenon, though to an exceedingly small extent, has just been made by Kundt and Röntgen in Germany.

Dr Kerr, of Glasgow, has extended the result to opaque bodies, and has shown that if light be passed through magnetised *iron* its plane is rotated. The film of iron must be exceedingly thin, because of its opacity; and hence, though the intrinsic rotating power of iron is undoubtedly very great, the observed rotation is exceedingly small and difficult to observe; and it is only by very remarkable patience and care and ingenuity that Dr Kerr has obtained his result. Mr Fitzgerald, of Dublin, has examined the question mathematically, and has shown that Maxwell's theory would have enabled Dr Kerr's result to be predicted.

Another requirement of the theory is that bodies which are transparent to light must be insulators or non-conductors of electricity, and that conductors of electricity are necessarily opaque to light. Simple observation amply confirms this. Metals are the best conductors, and are the most opaque bodies known. Insulators such as glass and crystals are transparent whenever they are sufficiently homogeneous, and the very remarkable researches of Professor Graham Bell in the last few months have shown that even *ebonite*, one of the most opaque insulators to ordinary vision, is certainly transparent to some kinds of radiation, and transparent to no small degree.

[The reason why transparent bodies must insulate, and why conductors must be opaque, was here illustrated by mechanical models.]

A further consequence of the theory is that the velocity of light in a transparent medium will be affected by its electrical strain constant; in other words, that its refractive index will bear some close but not yet quite ascertained relation to its specific inductive capacity. Experiment has partially confirmed this, but the confirmation is as yet very incomplete.

But there are a number of results not predicted by theory, and whose connection with the theory is not clearly made out. We have the fact that light falling on the platinum electrode of a voltameter generates a current, first observed, I think, by Sir W. R. Grove; at any rate it is mentioned in his 'Correlation of Forces'—extended by Becquerel and Robert Sabine to other substances, and now being extended to fluorescent and other bodies by Professor Minchin. And finally—for I must be brief—we have the remarkable action of light on selenium. This fact was discovered accidentally by an assistant in the laboratory of Mr Willoughby Smith, who noticed that a piece of selenium conducted electricity very much better when light was falling upon it than when it was in the dark. The light of a candle is sufficient, and instantaneously brings down the resistance to something like one-fifth of its original value.

This is the phenomenon which, as you know, has been utilised by Professor Graham Bell in that most ingenious and striking invention, the photophone.

I have now trespassed long enough upon your patience, but I must just allude to what may very likely be the next striking popular discovery, and that is the transmission of light by electricity. I mean the transmission of such things as views and pictures by means of the electric wire. It has not yet been done, but it seems already theoretically possible, and it may very soon be practically accomplished.

The Relation between Electricity and Light.

After Hertz.

Substance of a lecture by Prof. Oliver Lodge, Ashmolean Society, Oxford, June 3, 1889.[1]

For now wellnigh a century we have had a wave-theory of light; and a wave-theory of light is certainly true. It is directly demonstrable that lignt consists of waves of some kind or other, and that these waves travel at a certain well-known velocity, seven times the circumference of the earth per second, taking eight minutes on the journey from the sun to the earth. This propagation in time of an undulatory disturbance necessarily involves a medium. If waves setting out from the sun exist in space eight minutes before striking our eyes, there must necessarily be in space some medium in which they exist and which conveys them. Waves we cannot have unless they be waves in something.

[1] Based on a report in the (London) 'Electrician,' September 6, 1889.

No ordinary medium is competent to transmit waves at anything like the speed of light; hence the luminiferous medium must be a special kind of substance, and it is called the ether. The *luminiferous* ether it used to be called, because the conveyance of light was all it was then known to be capable of; but now that it is known to do a variety of other things also, the qualifying adjective may be dropped.

Wave motion in ether light certainly is; but what does one mean by the term wave? The popular notion is, I suppose, of something heaving up and down, or perhaps of something breaking on the shore in which it is possible to bathe. But if you ask a mathematician what he means by a wave, he will probably reply that the simplest wave is

$$y = a \sin (p\, t - n\, x),$$

and he might possibly refuse to give any other answer. And in refusing to give any other answer than this, or its equivalent in ordinary words, he is entirely justified; that *is* what is meant by the term wave, and nothing less general would be all-inclusive.

Translated into ordinary English, the phrase signifies "a disturbance periodic both in space and time." Anything thus doubly periodic is a wave; and all waves—whether in air as sound waves, or in ether as light waves, or on the surface of water as ocean waves—are comprehended in the definition.

What properties are essential to a medium capable of transmitting wave motion? Roughly we may say two—*elasticity* and *inertia*. Elasticity in some form, or some equivalent of it, in order to be able to store up energy and effect recoil; inertia, in order to enable the disturbed substance to overshoot the mark and oscillate beyond its place of equilibrium to and fro. Any medium possessing these two properties can transmit waves, and unless a medium possesses these properties in some form or other, or some equivalent for them, it may be said with moderate security to be incompetent to transmit waves. But if we make this latter statement one must be prepared to extend to the terms elasticity and inertia their very largest and broadest signification, so as to include any possible kind of restoring force and any possible kind of persistence of motion respectively.

These matters may be illustrated in many ways, but perhaps

a simple loaded lath or spring in a vice will serve well enough. Pull aside one end, and its elasticity tends to make it recoil; let it go, and its inertia causes it to overshoot its normal position: both causes together cause it to swing to and fro till its energy is exhausted. A regular series of such springs at equal intervals in space, set going at regular intervals of time one after the other, gives you at once a wave motion and appearance which the most casual observer must recognise as such. A series of pendulums will do just as well. Any wave-transmitting medium must similarly possess some form of elasticity and of inertia.

But now proceed to ask what is this ether which in the case of light is thus vibrating? What corresponds to the elastic displacement and recoil of the spring or pendulum? What corresponds to the inertia whereby it overshoots its mark? Do we know these properties in the ether in any other way?

The answer, given first by Clerk-Maxwell, and now reiterated and insisted on by experiments performed in every important laboratory in the world, is—

The elastic displacement corresponds to electro-static charge (roughly speaking, to electricity).

The inertia corresponds to magnetism.

This is the basis of the modern electro-magnetic theory of light. Now let me illustrate electrically how this can be.

The old and familiar operation of charging a Leyden jar—the storing up of energy in a strained dielectric—any electrostatic charging whatever—is quite analogous to the drawing aside of our flexible spring. It is making use of the elasticity of the ether to produce a tendency to recoil. Letting go the spring is analogous to permitting a discharge of the jar—permitting the strained dielectric to recover itself, the electrostatic disturbance to subside.

In nearly all the experiments of electro-statics ethereal elasticity is manifest.

Next consider inertia. How would one illustrate the fact that water, for instance, possesses inertia—the power of persisting in motion against obstacles—the power of possessing kinetic energy? The most direct way would be to take a stream of water and try suddenly to stop it. Open a water-tap freely and then suddenly shut it. The impetus or momentum of the stopped water makes itself manifest by a violent shock to the pipe, with which everybody must be familiar. The momentum of water is utilised by engineers in the "water-ram."

A precisely analogous experiment in electricity is what Faraday called "the extra current." Send a current through a coil of wire round a piece of iron, or take any other arrangement for developing powerful magnetism, and then suddenly stop the current by breaking the circuit. A violent flash occurs if the stoppage is sudden enough, a flash which means the bursting of the insulating air partition by the accumulated electro-magnetic momentum.

Briefly, we may say that nearly all electro-magnetic experiments illustrate the fact of ethereal inertia.

Now return to consider what happens when a charged conductor (say a Leyden jar) is discharged. The recoil of the strained dielectric causes a current, the inertia of this current causes it to overshoot the mark, and for an instant the charge of the jar is reversed: the current now flows backwards and charges the jar up as at first; again flows the current, and so on, discharging and charging the jar with rapid oscillations until the energy is all dissipated into heat. The operation is precisely analogous to the release of a strained spring, or to the plucking of a stretched string.

But the discharging body thus thrown into strong electrical vibration is embedded in the all-pervading ether, and we have just seen that the ether possesses the two properties requisite for the generation and transmission of waves—viz., elasticity, and inertia or density; hence, just as a tuning-fork vibrating in air excites aerial waves or sound, so a discharging Leyden jar in ether excites ethereal waves or light.

Ethereal waves can therefore be actually produced by direct electrical means. I discharge here a jar, and the room is for an instant filled with light. With light, I say, though you can see nothing. You can see and hear the spark indeed,—but that is a mere secondary disturbance we can for the present ignore—I do not mean any secondary disturbance. I mean the true ethereal waves emitted by the electric oscillation going on in the neighbourhood of this recoiling dielectric.

You pull aside the prong of a tuning-fork and let it go: vibration follows and sound is produced. You charge a Leyden jar and let it discharge: vibration follows and light is excited.

It is light just as good as any other light. It travels at the same pace, it is reflected and refracted according to the same laws; every experiment known to optics can be performed with this ethereal radiation electrically produced, and yet you cannot see it. Why not? For no fault of the light; the fault (if there be a fault) is in the eye. The retina is incompetent to respond to these vibrations—they are too slow. The vibrations set up when this large jar is discharged are from a hundred thousand to a million per second, but that is too slow for the retina. It responds only to vibrations between 4000 billions and 7000 billions per second. The vibrations are too quick for the ear, which responds only to vibrations between 40 and 40,000 per second. Between the highest audible and the lowest visible vibrations there has been hitherto a great gap, which these electric oscillations go far to fill up. There has been a great gap simply because we have no intermediate sense-organ to detect rates of vibration between 40,000 and 4,000,000,000,000,000 per second. It was, therefore, an unexplored territory. Waves have been there all the time in any quantity, but we have not thought about them nor attended to them.

It happens that I have myself succeeded in getting electric oscillations so slow as to be audible. The lowest I have got at present are 125 per second, and for some way above this the sparks emit a musical note; but no one has yet succeeded in directly making electric oscillations which are visible, though indirectly every one does it by lighting a candle.

Here, however, is an electric oscillator which vibrates 300 million times a second, and emits ethereal waves a yard long. The whole range of vibrations between musical tones and some thousand millions per second is now filled up.

These electro-magnetic waves have long been known on the side of theory, but interest in them has been immensely quickened by the discovery of a receiver or detector for them. The great though simple discovery by Hertz of an "electric eye,' as Sir W. Thomson calls it, makes experiments on these waves for the first time possible, or even easy. We have now a sort of artificial sense-organ for their appreciation—an electric arrangement which can virtually "see" these intermediate rates of vibration.

The Hertz receiver is the simplest thing in the world—nothing but a bit of wire, or a pair of bits of wire, adjusted so that when immersed in strong electric radiation they give minute sparks across a microscopic air-gap.

The receiver I have here is adapted for the yard-long waves emitted from this small oscillator; but for the far longer waves emitted by a discharging Leyden jar an excellent receiver is a gilt wall-paper or other interrupted metallic surface. The waves falling upon the metallic surface are reflected, and in the act of reflection excite electric currents, which cause sparks. Similarly, gigantic solar waves may produce auroræ; and minute waves from a candle do electrically disturb the retina.

The smaller waves are, however, far the most interesting and the most tractable to ordinary optical experiments. From a small oscillator, which may be a couple of small cylinders kept sparking into each other end to end by an induction coil, waves are emitted on which all manner of optical experiments can be performed.

They can be reflected by plain sheets of metal, concentrated by parabolic reflectors, refracted by prisms, concentrated by lenses. I have at the College a large lens of pitch, weighing over 3 cwt., for concentrating them to a focus. They can be made to show the phenomenon of interference, and thus have their wave lengths accurately measured. They are stopped by all conductors, and transmitted by all insulators. Metals are opaque, but even imperfect insulators, such as wood or stone, are strikingly transparent, and waves may be received in one room from a source in another, the door between the two being shut.

The real nature of metallic opacity and of transparency has long been clear in Maxwell's theory of light, and these electrically produced waves only illustrate and bring home the well-known facts. The experiments of Hertz are in fact the apotheosis of that theory.

Thus, then, in every way Maxwell's brilliant perception of the real nature of light is abundantly justified; and for the first time we have a true theory of light, no longer based upon analogy with sound, nor upon a hypothetical jelly or elastic solid.

Light is an electro-magnetic disturbance of the ether. Optics is a branch of electricity. Outstanding problems in optics are being rapidly solved now that we have the means of definitely exciting light with a full perception of what we are doing, and of the precise mode of its vibration.

It remains to find out how to shorten down the waves—to hurry up the vibration until the light becomes visible. Nothing is wanted but quicker modes of vibration. Smaller oscillators must be used—very much smaller—oscillators not much bigger than molecules. In all probability—one may almost say certainly—ordinary light is the result of electric oscillation in the molecules of hot bodies, or sometimes of bodies not hot—as in the phenomenon of phosphorescence.

The direct generation of *visible* light by electric means, so soon as we have learnt how to attain the necessary frequency of vibration, will have most important practical consequences.

For consider our present methods of making artificial light: they are both wasteful and ineffective.

We want a certain range of oscillation, between 7000 and 4000 billion vibrations per second,—no other is useful to us, because no other has any effect upon our retina; but we do not know how to produce vibrations of this rate. We can produce a definite vibration of one or two hundred or thousand per second—in other words, we can excite a pure tone of definite pitch; and we can command any desired range of such tones continuously by means of bellows and a keyboard. We can also (though the fact is less well known) excite momentarily definite ethereal vibrations of some millions per second, as I have explained; but we do not at present seem to know how to maintain this rate quite continuously. To get much faster rates of vibration than this, we have to fall back upon atoms. We know how to make atoms vibrate,—it is done by what we call "heating" the substance; and if we could deal with individual atoms unhampered by others, it is possible that we might get a pure and simple mode of vibration from them. It is possible, but unlikely; for atoms, even when isolated, have a multitude of modes of vibration special to themselves, of which only a few are of practical use to us, and we do not know how to excite some without also the others. However, we do not at present even deal with individual atoms; we treat them crowded together in a compact mass, so that their modes of vibration are really infinite.

We take a lump of matter, say a carbon filament or a piece of quicklime, and by raising its temperature we impress upon its atoms higher and higher modes of vibration, not transmuting the lower into the higher, but superposing the higher upon the lower, until at length we get such rates of vibration as our retina is constructed for, and we are satisfied. But how wasteful and indirect and empirical is the process! We want a small range of rapid vibrations, and we know no better than to make the whole series leading up to them. It is as though, in order to sound some little shrill octave of pipes in an organ, we are obliged to depress every key and every pedal, and to blow a young hurricane.

I have purposely selected as examples the more perfect methods of obtaining artificial light, wherein the waste radiation is only useless, and not noxious. But the old-fashioned plan was cruder even than this: it consisted simply in setting something burning, whereby not the fuel but the air was consumed; whereby also a most powerful radiation was produced, in the waste waves of which we were content to sit stewing, for the sake of the minute—almost infinitesimal—fraction of it which enabled us to see.

Every one knows now, however, that combustion is not a pleasant or healthy mode of obtaining light; but everybody does not realise that neither is incandescence a satisfactory and unwasteful method, which is likely to be practised for more than a few decades, or perhaps a century.

Look at the furnaces and boilers of a great steam-engine driving a group of dynamos, and estimate the energy expended; and then look at the incandescent filaments of the lamps excited by them, and estimate how much of their radiated energy is of real service to the eye. It will be as the energy of a pitch-pipe to an entire orchestra.

It is not too much to say that a boy turning a handle could, if his energy were properly directed, produce quite as much real light as is produced by all this mass of mechanism and consumption of material. There might, perhaps, be something contrary to the laws of nature in thus hoping to get and utilise some specific kind of radiation without the rest; but Lord Rayleigh has shown in a short communication to the British Association at York that it is not so, and that therefore we have a right to try to do it.

We do not yet know how, it is true, but it is one of the things we have got to learn.

Any one looking at a common glowworm must be struck with the fact that not by ordinary combustion, nor yet on the steam-engine and dynamo principle, is that easy light produced. Very little waste radiation is there from phosphorescent things in general. Light of the kind able to affect the retina is directly emitted; and for this, for even a large supply of this, a modicum of energy suffices.

Solar radiation consists of waves of all sizes, it is true; but then solar radiation has innumerable things to do besides making things visible. The whole of its energy is useful. In artificial lighting nothing but light is desired; when heat is wanted it is best obtained separately by combustion. And so soon as we clearly recognise that light is an electrical vibration, so soon shall we begin to beat about for some mode of exciting and maintaining an electrical vibration of any required degree of rapidity. When this has been accomplished, the problem of artificial lighting will have been solved.

APPENDIX B.

PROF. HENRY ON HIGH TENSION ELECTRICITY BEING CONFINED TO THE SURFACE OF CONDUCTING BODIES, WITH SPECIAL REFERENCE TO THE PROPER CONSTRUCTION OF LIGHTNING-RODS.

(*Extracted from the* '*Journal of the Telegraph,*' *New York, Sept.* 1, 1877.)

WASHINGTON, *March* 11, 1876.

DEAR SIR,—In answer to your letter of the 7th inst., I have to say that the discrepancy which exists as to the question whether electricity passes at the surface or through the whole capacity of the rod has arisen principally from experiments on galvanic electricity, which, having little or no repulsive energy, passes through the whole substance of the rod, and also from experiments in which a very large quantity of frictional electricity is transmitted through a small wire: in this case the metal is resolved into its elements and reduced to an impalpable powder.

In the case, however, of the transmission of atmospheric electricity through a rod of sufficient size to transmit the discharge freely, there can be no doubt that it tends to pass at the surface, the thickness of the stratum of electricity varying with the diameter of the rod and the amount and the intensity of the charge.

To test this by actual experiment I made the following arrangement: through a gun-barrel about 2 feet in length a copper wire was passed, the ends projecting. The middle of the wire in the barrel was coiled into the form of a magnetising spiral, and the ends of the gun-barrel were closed with plugs of tinfoil, so as to make a perfect metallic connection between the wire and the barrel. On the outside of the barrel another magnetising spiral was placed, the whole arrangement being shown in the sketch.

A powerful charge was now sent through the copper wire from a Leyden jar of about two gallons' capacity. The needle within the barrel showed not the least sign of magnetism, while the one on the outside was strongly magnetic.

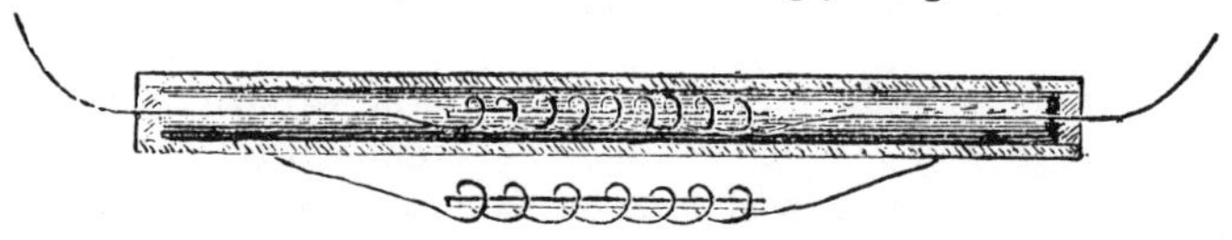

From this experiment I conclude that a gas-pipe can convey an ordinary charge of electricity from the clouds as well as a solid rod of the same diameter.

The repulsive energy of the electrical discharge at right angles to the axis remains of the same intensity as in the case of a statical charge. This I have shown to be the case by drawing sparks of considerable intensity from a conductor, one end of which was connected with the ground while sparks were thrown on the other end from a large prime conductor. This spark is of a peculiar character, for though it gives a pungent shock and sets fire to combustible substances, such as an electrical pistol, it does not affect a sensitive gold-leaf electrometer. The fact is, it consists of two sparks, the one negative and the other positive. The rod during the transmission of the electricity through it is charged + at the upper end, and immediately in advance of this point it is charged − by induction, and the electricity passes through it in the discharge in the form of a series of + and − waves.—Yours very truly, JOSEPH HENRY, *Sec. Smithsonian Inst.*

Prof. R. C. KEDZIE, Lansing, Michigan.

WASHINGTON, *April* 15, 1876.

DEAR SIR,—Your letter was received by due course of mail, but a press of business connected with the preparation of the Annual Report for 1875 and the Lighthouse Board has prevented an earlier reply.

I have now to say that, as far as I know, I am the only person who has made a special study of the conduction of frictional electricity in regard to lightning-rods. It has long been established by Coulomb and others that the electricity of a charged conductor exists in a thin stratum at the surface, and this is a necessary consequence of the repulsion of electricity for itself, every particle being repelled from every other as far as possible. From this it was hastily assumed that electricity in motion also moves at the surface; but this was an inference without physical proof until I commenced the investigation. I found from a series of experiments that *frictional* electricity—that is, electricity of repulsive energy, such as that from the clouds—does pass at the surface, but that *galvanic* electricity, the kind to which Faraday, Daniell, De La Rive, and others refer, passes through the whole capacity of the conductor. This latter fact, however, was previously established by others. I further found that whenever a charge of electricity was thrown on a rod explosively, however well connected the rod was with the earth, it gave off sparks in the course of its length sufficient to fire an electric pistol and light flocculent substances. I also found that, in sending a powerful discharge from a battery of nine jars through a wide plate, no electricity passed along the middle of the plate, but that it was accumulated in its passage at the edges.

From all my study of this subject I do not hesitate to say that the plan I have given of lightning-rods is the true one, and that a tube of a sufficient degree of thickness serves to conduct the electricity as well as a solid mass, provided the thickness is sufficient to give free conduction. A very heavy charge sent through a wire frequently deflagrates it, but no discharge from the clouds, of which I have any knowledge, has ever sufficed to deflagrate a gas-pipe of an inch in diameter.

The plan of increasing the surface of a rod by converting the metal into a ribbon is objectionable. It tends to increase the power of the lateral discharge, and gives no increase of conducting power.

Another fallacy is much insisted on—viz., the better conduction of copper than iron. It is true that copper is a better conductor of *galvanic* electricity, which pervades the whole mass, but in regard to *frictional* electricity the difference in conducting capacity is too small to be of any importance. Iron is sufficiently good in regard to conduction, and withstands deflagration better than copper: besides this, it is much cheaper.—Yours truly, JOSEPH HENRY.

Prof. R. C. KEDZIE.

On Modern Views with Respect to the Nature of Electric Currents.

Substance of a lecture by Prof. H. A. Rowland, American Institute of Electrical Engineers, May 22, 1889.[1]

. . . .

How great, then, the difference between a current of water and a current of electricity! The action of the former is confined to the interior of the tube, while that of the latter extends to great distances on all sides, the whole of the space being agitated by the formation of an electric current in any part. To show this agitation, I have here two large frames with coils of wire around them. They hang face to face about 6 feet apart. Through one I discharge this Leyden jar, and immediately you see a spark at a break in the wire of the other coil, and yet there is no apparent connection between the two. I can carry the coils 50 feet or more apart, and yet, by suitable means, I can observe the disturbances due to the current in the first coil.

The question is forced upon us as to how this action takes place. How is it possible to transmit so much power to such a distance across apparently unoccupied space? According to our modern theories of physics, there must be some medium engaged in this transmission. We know that it is not the air, because the same effects take place in a vacuum, and therefore we must fall back on that medium which transmits light, and which we have named the ether—that medium which is supposed to extend unaltered throughout the whole of space, whose existence is very certain, but whose properties we have yet but vaguely conceived.

I cannot in the course of one short hour give even an idea of the process by which the minds of physicists have been led to this conclusion, or the means by which we have finally completely identified the ether which transmits light with the medium which transmits electrical and magnetic disturbances. The great genius who first identified the two is Maxwell, whose electro-magnetic theory of light is the centre around

[1] Based on reports in the (London) 'Electrician,' June 21 and 28 1889.

which much scientific thought is to-day revolving, and which we regard as one of the greatest steps by which we advance nearer to the understanding of matter and its laws. It is this great discovery of Maxwell which allows me to attempt to explain to you the wonderful events which happen everywhere in space when one establishes an electric current in any other portion.

In the first place, we discover that the disturbance does not take place in all portions of space at once, but proceeds outward from the centre of the disturbance with a velocity exactly equal to the velocity of light; so that when I touch these wires together so as to complete the circuit of yonder battery, I start a wave of ethereal disturbance which passes outward with a velocity of 185,000 miles per second, and continues to pass outwards for ever, or until it reaches the bounds of the universe. And yet none of our senses informs us of what has taken place unless sharpened by the use of suitable instruments. Thus, in the case of these two coils of wire, suspended near each other, when the wave from the primary disturbance reaches the second coil we perceive the disturbance by means of the spark formed at the break in the coil. Should I move the coils farther apart, the spark in the second coil would be somewhat delayed, but the distance of 185,000 miles would be necessary before this delay could amount to as much as one second. Hence the effects we observe on the earth take place so nearly instantaneously that the interval of time is very difficult to measure, amounting in the present case to only $\frac{1}{1500000000}$th of a second.

It is impossible for me to prove the existence of this interval, so infinitesimal is it, but I can at least show you that waves have something to do with the action observed. For instance, I have here two tuning-forks mounted on sounding-boxes and tuned to exact unison. I sound one and then stop its vibrations with my hand; instantly you hear that the other is in vibration, caused by the waves of sound in the air between the two. When, however, I destroy the unison by fixing this piece of wax on one of the forks, the action ceases.

Now, this combination of a coil of wire and a Leyden jar forms a vibrating system of electricity, and its time of vibration is about 10,000,000 times a second. Here is another combination of coil and jar, the same as the first, and therefore its time of vibration is the same. You see how well the experiment works, because the two are in unison. But let me take away this second Leyden jar, thus destroying the unison, and you see that the sparks instantly cease. Replacing it, the sparks reappear. Adding another on one side, they disappear again, only to reappear when the system is made symmetrical by placing two on each side.

This experiment and that of the tuning-forks have an exact analogy to one another. In each we have two vibrating systems connected by a medium capable of transmitting vibrations, and they both come under the head of what we know as sympathetic vibrations. In the one case, we have two mechanical tuning-forks connected by the air; in the other, two pieces of apparatus, which we might call electrical tuning-forks, connected by the ether. The vibrations in one case can be seen by the eye or heard by the ear, but in the other case they can only be perceived when we destroy them by making them produce a spark. The fact that we are able to increase the effect by proper tuning demonstrates that vibrations are concerned in the phenomenon. This can, however, be separately demonstrated by examining the spark by means of a revolving mirror, when we find that it is made up of many successive sparks corresponding to the successive backward and forward movements of the current.

Thus, in the case of a charged Leyden jar whose inner and outer coatings have been suddenly joined by a wire, the electricity flows back and forth along the wire until all the energy originally stored up in the jar has expended itself in heating the wire or the air where the spark takes place, and in generating waves of disturbance in the ether which move outward into space with the velocity of light. These ethereal waves we have demonstrated by letting them fall on this coil of wire, causing the electrical disturbance to manifest itself by electric sparks.

I have here another more powerful arrangement for producing electro-magnetic waves of very long wave length, each one being about 500 miles long. It consists of a coil within which is a bundle of iron wires. On passing a powerful alternating current through the coil the iron wires are rapidly magnetised and demagnetised, and send forth into space a system of electro-magnetic waves at the rate of 360 in a second.

Here also I have another piece of apparatus for sending out the same kind of electro-magnetic waves, and on applying a match we start it also into action. But the last apparatus is tuned to so high a pitch that the waves are only $\frac{1}{50000}$ inch long, and 55,000,000,000,000 are given out in one second. These short waves are known by the name of light and radiant heat, though the name radiation is more exact. Placing any body near the lamp so that the radiation can fall on it, we observe that when the body absorbs the rays it is heated by them. Is it not possible for us to get some substance to absorb the long (or electro-magnetic) waves of disturbance, and so obtain a heating effect? I have here such a substance in the shape of a sheet of copper, which I fasten on the face of a thermopile, and I hold it where these waves are strongest. As I have anticipated, great heat is generated by their absorption, and soon the plate of copper becomes very warm, as we see by this thermometer, by feeling it with the hand, or even by the steam from water thrown upon it. In this experiment the copper had not touched the coil or the iron wire core, although if it did they are very much cooler than itself. The heat has been produced by the absorption of the waves in the same way as a blackened body absorbs the rays of shorter wave length from the lamp.

In these experiments, so far, the wave-like nature of the disturbance has not been proved. We have caused electric sparks, and have heated the copper plate across an interval of space, but have not in either of these cases proved experimentally the progressive nature of the disturbance.

A ready means of experimenting on the waves, obtaining their wave length and showing their interferences, has hitherto been wanting. This deficiency has been recently supplied by Prof. Hertz, of Carlsruhe.

I scarcely know how to present this subject to a non-technical audience and make it clear how a coil of wire with a break in it can be used to measure the velocity and length of ethereal waves. However, I can but try. If the waves moved very slowly, we could readily measure the time the first coil took to affect the second, and show that this time was longer as the distance was greater. But it is absolutely inappreciable by any of our instruments, and another method must be found. To obtain the wave length Prof. Hertz used several methods, but that by the formation of stationary waves is the most easily grasped. I hold in my hand one end of a spiral spring, which makes a heavy and flexible rope. As I send a wave down it, you see that it is reflected at the farther end, and returns again to my hand. If, however, I send a succession of waves down the rope, the reflected waves interfere with the direct ones, and divide the rope into a succession of nodes and loops which you now observe. So, a series of sound waves, striking on a wall, forms a system of stationary waves in front of the wall. Indeed we can use any waves for this purpose, even ethereal waves. With this in view Prof. Hertz established his apparatus in front of a reflecting wall, and observed the nodes and loops by the sparks produced in a ring of wire, somewhat resembling the coil I have been using, but much smaller. It is impossible for me to repeat this experiment before you, as it is a very delicate one, and the sparks produced are almost microscopic. Indeed I should have to erect an entirely different apparatus, as the waves from the one before me are nearly a quarter-mile long. To produce shorter waves we must use apparatus very much smaller—tuned, as it were, to a higher pitch, so that several stationary waves, or nodes and loops, of a few yards long could be obtained in the space of this room.

The testing coil would then be moved to different parts of the room, and the nodes would be indicated by the disappearance of the sparks, and the loops by the greater brightness of them. The presence of stationary waves would thus be proved, and their half-wave length found from the distance from node to node, for stationary waves can always be considered as produced by the interference of two waves advancing in opposite directions.

The closing of a battery circuit, then, and the establishment of a current of electricity in a wire, is a very different process from the formation of a current of water in a pipe, though after the first shock the laws of the flow of the two are very much alike. Furthermore, the medium around the current of electricity has very strange properties, showing that it is accompanied by a disturbance throughout space. The wire is but the core of the disturbance, which latter extends indefinitely in all directions.

One of the strangest things about it is that we can calculate with perfect exactness the velocity of the wave propagation and the amount of the disturbance at every point and at any instant of time ; but as yet we cannot conceive of the details of the mechanism which is concerned in the propagation of an electric current. In this respect our subject resembles all other branches of physics in the partial knowledge we have of it. We know that light is the undulation of the luminiferous ether, and yet the constitution of the latter is unknown. We know that the atoms of matter can vibrate with purer tones than the most perfect piano, and yet we cannot even conceive of their constitution. We know that the sun attracts the planets with a force whose law is known, and yet we fail to picture to ourselves the process by which it takes our earth within its grasp at the distance of many millions of miles and prevents it from departing for ever from its life-giving rays. Science is full of this half-knowledge.

So far we have considered the case of alternating electric currents in a wire connecting the inner and outer coatings of a Leyden jar. The invention of the telephone, by which sound is carried from one point to another by means of electrical waves, has forced into prominence the subject of these waves. Furthermore, the use of alternating currents for electric lighting brings into play the same phenomenon. Here, again, the difference between a current of water and a current of electricity is very marked. A sound wave, traversing the water in the tube, produces a to-and-fro current of water at any given point. So, in the electrical vibration along a wire, the electricity moves to and fro along it in a manner somewhat similar to the water, but with this difference : the disturbance from the water-motion is confined to the tube, and the oscillation of the water is greatest in the centre of the tube ; while in the case of the electric current the ether around the wire is disturbed, and the oscillation of the current is greatest at the surface of the wire and least in its centre. The oscillations in the water take place in the tube without reference to the matter outside the tube, whereas the electric oscillations in the wire are entirely dependent on the surrounding space, and the velocity of the propagation is nearly independent of the nature of the wire, provided it is a good conductor.

We have then in the case of electrical waves along a wire a disturbance outside the wire and a current within it, and the equations of Maxwell allow us to calculate these with perfect accuracy and give all the laws with respect to them.

We thus find that the velocity of propagation of the waves along a wire, hung far away from other bodies and made of good conducting material, is that of light, or 185,000 miles per second ; but when it is hung near any conducting matter, like the earth, or enclosed in a cable and sunk into the sea, the velocity becomes much less. When hung in space, away from other bodies, it forms, as it were, the core of a system of waves in the ether, the amplitude of the disturbance becoming less and less as we move away from the wire. But the most curious fact is that the electric current penetrates only a short distance into the wire, being mostly confined to the surface, especially where the number of oscillations per second is very great.

The electrical waves at the surface of a conductor are thus, in some respects, very similar to the waves on the surface of water. The greatest motion in the latter case is at the surface, while it diminishes as we pass downwards and soon becomes inappreciable. Furthermore, the depth to which the disturbance penetrates into the water increases with increase of the length of the wave, being confined to very near the surface for very short waves. So the disturbance in the copper penetrates deeper as the waves and the time of oscillation are longer, and the disturbance is more nearly confined to the surface as the waves become shorter.[1]

There are very many practical applications of these theoretical results for electric currents. The most obvious one is to the case of conductors for the alternating currents used

[1] A striking illustration of this skin-deep penetration of high-voltage electricity was communicated by Lord Armstrong to Sir William Thomson (now Lord Kelvin) at the Newcastle meeting of the British Association in 1889. A bar of steel about a foot long, which Lord Armstrong was holding in his hand, was allowed accidentally to short circuit the two terminals of a dynamo giving an alternate current of 85 ampères, at a difference of potential of 103 volts. He instantly felt a sensation of

in producing the electric light. We find that when these are larger than about half an inch diameter they should be replaced by a number of conductors less than half an inch diameter, or by strips about a quarter of an inch thick, and of any convenient width.

Prof. Oliver Lodge has recently drawn attention to another application of these results—that is, to lightning-rods. Almost since the time of Franklin there have been those who advocated the making of lightning-rods hollow in order to increase the surface for a given amount of copper. We now know that these persons had no reason for their belief, as they simply drew the inference that electricity at rest is on the surface. Neither were the advocates of the solid rods quite correct, for they reasoned that electricity in a state of steady flow occupies the whole area of the conductor equally. The true theory, we now know, indicates that neither party was entirely correct, and that the surface is a very important factor in the case of a current of electricity so sudden as that from a lightning discharge. But increase of surface can best be obtained by multiplying the number of conductors, rather than making them flat or hollow. Theory indicates that the current penetrates only one-tenth the distance into iron that it does into copper. As the iron has seven times the resistance of copper, we should need seventy times the surface of iron that we should of copper. Hence I prefer copper wire about a quarter of an inch diameter and nailed directly to the house without insulators, and passing down the four corners, around the eaves, and over the roof, for giving protection from lightning in all cases where a metal roof and metal down-spouts do not accomplish the same purpose.

Whether the discharge of lightning is oscillatory or not does not enter into the question, provided it is only sufficiently sudden. I have recently solved the mathematical problem of the electric oscillations along a perfectly conducting wire join-

burning and dropped the bar. His fingers were badly blistered, though on examining the bar a few seconds afterwards it was found to be *quite cold*. This proved that the action lay at the surface, and had not time to sensibly penetrate the substance of the bar. There were two little hollows burned out of the metal at the points where it touched the dynamo terminals.—J. J. F.

ing two infinite and perfectly conducting planes parallel to each other, and find that there is no definite time of oscillation, but that the system is capable of vibrating in any time in which it is originally started. The case of lightning between a cloud of limited extent and the earth along a path through the air of great resistance is a very different problem. Both the cloud and the path of the electricity are poor conductors, which tends to lengthen the time. If I were called on to estimate as nearly as possible what took place in a flash of lightning, I would say that I did not believe that the discharge was always oscillating, but more often consisted of one or more streams of electricity at intervals of a small fraction of a second, each one continuing for not less than $\frac{1}{100000}$ second. An oscillating current with 100,000 reversals per second would penetrate about $\frac{1}{60}$ inch into copper and $\frac{1}{600}$ inch into iron. The depth for copper would constitute a considerable proportion of a wire $\frac{1}{4}$ inch diameter, and as there are other considerations to be taken into account, I believe it is scarcely worth while making tubes, or flat strips, for such small sizes.

It is almost impossible to draw proper conclusions from experiments on this subject in the laboratory, such as those of Prof. Oliver Lodge.[1] The time of oscillation of the current in most pieces of laboratory apparatus is so very small, being often the $\frac{1}{100000000}$ of a second, that entirely wrong inferences may be drawn from them. As the size of the apparatus increases, the time of oscillation increases in the same proportion, and changes the whole aspect of the case. I have given $\frac{1}{100000}$ of a second as the shortest time a lightning-flash could probably occupy. I strongly suspect it is often much greater, and thus departs even further from the laboratory experiments of Prof. Lodge, who has, however, done very much towards drawing attention to this matter and showing the importance of surface in this case. All shapes of the rod with equal surface are not, however, equally efficient. Thus, the inside surface of a tube does not count at all. Neither do the corrugations on a rod count for the full value of the surface they

[1] For Prof. Lodge's views see his paper, 'Jour. Inst. Elec. Engs.,' vol. xix. p. 352, and the very interesting discussion thereupon.—J. J. F.

expose, for the current is not distributed uniformly over the surface; but I have recently proved that rapidly alternating currents are distributed over the surface of very good conductors in the same manner as electricity at rest would be distributed over them, so that the exterior angles and corners possess much more than their share of the current, and corrugations on the wire concentrate the current on the outer angles and diminish it in the hollows. Even a flat strip has more current on the edges than in the centre.

For these reasons, shape, as well as extent of surface, must be taken into account, and strips have not always an advantage over wires for quick discharges.

The fact that the lightning-rod is not melted on being struck by lightning is not now considered as any proof that it has done its work properly. It must, as it were, seize upon the discharge, and offer it an easier passage to the earth than any other. Such sudden currents of electricity we have seen to obey very different laws from continuous ones, and their tendency to stick to a conductor and not fly off to other objects depends not only on having them of small resistance, but also on having what we call the self-induction as small as possible. This latter can be diminished by having the lightning-rod spread sideways as much as possible, either by rolling it into strips, or better, by making a network of rods over the roof with several connections to the earth at the corners, as I have before described.

Thus we see that the theory of lightning-rods, which appeared so simple in the time of Franklin, is to-day a very complicated one, and requires for its solution a very complete knowledge of the dynamics of electric currents. In the light of our present knowledge the frequent failure of the old system of rods is no mystery, for I doubt if there are a hundred buildings in the country properly protected from lightning. With our modern advances, perfect protection might be guaranteed in all cases, if expense were no object.

We have now considered the case of oscillations of electricity in a few cases, and can turn to that of steady currents. The closing of an electric circuit sends ethereal waves throughout space, but after the first shock the current flows steadily without producing any more waves. However, the properties of the space around the wire have been permanently altered, as we have already seen. Let us now study these properties more in detail. I have before me a wire in which I can produce a powerful current of electricity, and we have seen that the space around it has been so altered that a delicately suspended magnetic needle cannot remain quiet in all positions, but stretches itself at right angles to the wire, the north pole tending to revolve around it in one direction and the south pole in the other. This is a very old experiment, but we now regard it as evidence that the properties of the space around the wire have been altered rather than that the wire acts on the magnet from a distance.

Put, now, a plate of glass around the wire, the latter being vertical and the former with its plane horizontal, and pass a powerful current through the wire. On now sprinkling iron filings on the plate they arrange themselves in circles around the wire, and thus point out to us the celebrated lines of magnetic force of Faraday. Using two wires with currents in the same direction we get these other curves, and, testing the forces acting on the wire, we find that they are trying to move towards each other.

Again, pass the currents in the opposite directions and we get these other curves, and the currents repel each other. If we assume that the lines of force are like rubber bands which tend to shorten in the direction of their length and repel each other sideways, Faraday and Maxwell have shown that all magnetic attractions and repulsions are explained. The property which the presence of the electric current has conferred on the ether is then one by which it tends to shorten in one direction and spread out in the other two directions.

We have thus done away with action at a distance, and have accounted for magnetic attraction by a change in the intervening medium, as Faraday partly did almost fifty years ago. For this change in the surrounding medium is as much a part of the electric current as anything that goes on within the wire.

To illustrate this tension along the lines of force, I have constructed this model, which represents the section of a coil of wire with a bar of iron within it. The rubber bands represent the lines of force which pass around the coil and through the iron bar, as they have an easier passage through the iron than the air. As we draw the bar down and let it go, you see that it is drawn upward and oscillates around its position of equilibrium until friction brings it to rest. Here, again, I have a coil of wire with an iron bar within it with one end resting on the floor. As we pass the current, and the lines of magnetic force form around the coil and pass through the iron, it is lifted upwards, although it weighs 24 lb., and oscillates around its position of equilibrium exactly the same as though it were sustained by rubber bands as in the model. The rubber bands in this case are invisible to our eye, but our mental vision pictures them as lines of magnetic force in the ether drawing the bar upward by their contractile force. This contractile force is no small quantity, as it may amount, in some cases, to one or even two hundred pounds to the square inch, and thus rivals the greatest pressure which we use in our steam-engines.

Thus the ether is, to-day, a much more important factor in science than the air we breathe. We are constantly surrounded by the two, and the presence of the air is manifest to us all; we feel it, we hear by its aid, and we even see it under favourable circumstances, and the velocity of its motion as well as the amount of moisture it carries is a constant topic of conversation. The ether, on the other hand, eludes all our senses, and it is only with imagination, the eye of the mind, that its presence can be perceived. By its aid in conveying the vibrations we call light we are enabled to see the world around us; and by its other motions, which cause magnetism, the mariner steers his ship through the darkest night when the heavenly bodies are hid from view. When we speak in a telephone, the vibrations of the voice are carried forward to the distant point by waves in the ether, there again to be resolved into the sound waves of the air. When we use the electric light to illuminate our streets, it is the ether which conveys the energy along the wires as well as transmits it to our eye after it has assumed the form of light. We step upon an electric street-car and feel it driven forward with the power of many horses, and again it is the ether whose immense force we have brought under our control and made to serve our purpose—no longer a feeble, uncertain sort of medium, but a mighty power, extending throughout all space, and binding the whole universe together.

APPENDIX C.

VARIATIONS OF CONDUCTIVITY UNDER ELECTRICAL INFLUENCE.

Substance of a paper by Prof. E. Branly, of the Catholic University of Paris.[1]

The object of this article is to describe the first results obtained in an investigation of the variation or resistance of a large number of conductors under various electrical influences. The substances which up to the present have presented the greatest variations in conductivity are the powders or filings of metals. The enormous resistance offered by metal in a state of powder is well known ; indeed, if we take a somewhat long column of very fine metallic powder, the passage of the current is completely stopped. The increase in the electrical conductivity by pressure of powdered conducting substances is also well known, and has had various practical applications. The variations of conductivity, however, which occur on subjecting such bodies to various electrical influences have not been previously investigated.

The Effect of Electric Sparks.—Let us take a circuit comprising a single cell, a galvanometer, and some powdered metal enclosed in an ebonite tube of one square centimetre cross section and a few centimetres long. Close the extremities of the tube with two cylindrical copper tubes pressing against the powdered metal and connected to the rest of the circuit. If the powder is sufficiently fine, even a very sensitive galvanometer does not show any evidence of a current passing. The resistance is of the order of millions of ohms, although the same metal melted or under pressure would only offer (the dimensions being the same) a resistance equal to a fraction of an ohm. There being, therefore, no current in the circuit, a Leyden

[1] Based on reports in the (London) 'Electrician,' June 26 and August 21, 1891.

jar is discharged at some little distance off, when the abrupt and permanent deflection of the galvanometer needle shows that an immediate and a permanent reduction of the resistance has been caused. The resistance of the metal is no longer to be measured in millions of ohms, but in hundreds. Its conductivity increases with the number and intensity of the sparks.

Some 20 or 30 centimetres from a circuit comprising some metallic filings contained in an ebonite cup, let us place a hollow brass sphere, 15 to 20 centimetres in diameter, insulated by a vertical glass support. The filings offer an enormous resistance and the galvanometer needle remains at zero. But if we bring an electrified stick of resin near the sphere, a little spark will pass between the stick and the sphere, and immediately the needle of the galvanometer is violently jerked and then remains permanently deflected. On some fresh filings being placed in the ebonite cup, the resistance of the circuit will again keep the needle at zero. If now the charged brass sphere is touched with the finger, there is a minute discharge and the galvanometer needle is again deflected. With a few accumulators the experiment can easily be made without a galvanometer. The circuit consists of the battery, some metallic powder, a platinum wire, and a mercury cup. The resistance of the powder is so high that the interruption of the circuit takes place without any sparking of the mercury cup. If now a Leyden jar is discharged in the neighbourhood of the circuit the powder is rendered conducting, the platinum wire immediately becomes red hot, and a violent spark occurs on breaking the circuit.

The influence of the spark decreases as the distance increases, but its influence is observable several metres away from the powder, even with a small Wimshurst machine. Repeating the spark increases the conductivity; in fact, with certain substances successive sparks produce successive jerks, and a gradually increasing and persistent deflection of the galvanometer needle.

Influence of a Conductor traversed by Condenser Discharges.— While using the Wimshurst machine it was noticed that the reduction in the resistance of the filings frequently took place before discharge. This led me to the following experiment:

Take a long brass tube, one end of which is close to the circuit containing the metallic powder ; its other end, several metres distant from the circuit, is fairly close to a charged Leyden jar. A spark takes place and the conductor is charged. At the same instant, the conductivity of the metallic powder is greatly increased.

The following arrangement, owing to its efficacy, convenience, and regularity of action, was used by me in most of my researches, and I shall briefly call it the A arrangement (fig. 1).

The source of electricity is a two-plate Holtz machine driven at from 100 to 400 revolutions. A sensitive substance is introduced into one of the arms of a Wheatstone bridge, or into the circuit of a single Daniell cell at a distance of some 10 metres from the Holtz machine. Between the discharge knobs of the machine and the Wheatstone bridge, and connected to the former, there are two insulated brass tubes, A A′, running parallel to one another 40 centimetres apart. The Leyden jars usually attached to a Holtz machine may be dispensed with, the capacity of the long brass tubes being in some measure equivalent to them. The knobs S were 1 mm., ·5 mm., or ·1 mm. apart. When the plates were rotated, sparks rapidly succeeded each other. Experiments showed that these sparks had no direct effect at a distance of 10 metres. The two tubes A A′ are not absolutely necessary ; the diminution of resistance is easily produced if only one is employed, and in some cases, indeed, a single conductor is more efficacious. An increase in the speed of the machine increases its action to a marked extent. The sparks at S may be suppressed by drawing the knobs apart, but the conductor A will still continue to exert its influence, especially if there is a spark-gap anywhere about.

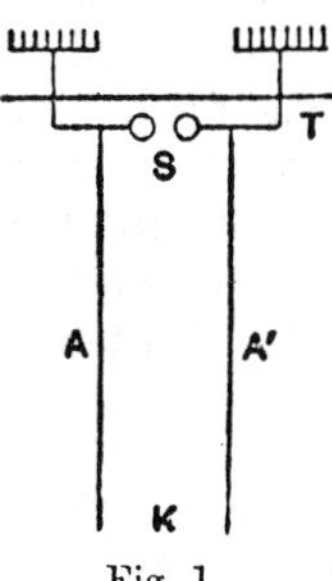

Fig. 1.

Effects of Induced Currents.—The passage of induced currents *through* a sensitive substance produces similar effects to those described above. In one instance an induction coil was taken, having two similar wires. The circuit of the secondary wire was closed through a tube containing filings, the galvanometer being also in circuit. Care was taken to ascertain before introducing the filings into the circuit that the currents on make and break gave equal and opposite deflections. Filings were then introduced into the circuit, the primary being made and broken at regular intervals. The following table gives the results obtained in the case of zinc filings :—

ZINC FILINGS.

Galvanometer throws.		Galvanometer throws.	
1st closing . .	1°	1st opening . .	18°
2nd ,, . .	64°	2nd ,, . .	100°
3rd ,, . .	146°	3rd ,, . .	140°

Effects of passing Continuous Currents of High E.M.F.—If a continuous current of high E.M.F. is employed, it renders a sensitive substance conducting. The phenomenon may be shown in the following manner. A circuit is made up consisting of a battery, a sensitive substance, and a galvanometer. The E.M.F. of the battery is first 1 volt, then 100 volts, then 1 volt. Below I give the galvanometer deflections obtained with an E.M.F. of 1 volt for three different substances before and after the application of the E.M.F. of 100 volts :—

Before application of current.	After application of current.
16	100
0	15
1	500

In the case of some measurements taken on a Wheatstone bridge, a prism of aluminium filings interposed between two copper electrodes offered a resistance of several million ohms before a high E.M.F. was applied, but only offered a resistance of 350 ohms after the application of this pressure for one minute. The time during which the powder should be interposed in the battery circuit should not be too short. Thus, in one instance the application for 10 seconds of 75 mercury sulphate cells produced no effect, but their application for 60 seconds resulted in the resistance being reduced from several megohms to 2500 ohms.

It should be observed that the phenomenon of suddenly increased conductivity occurs even if the sensitive substance is not in circuit with a battery at the time it is influenced. Thus, the metallic filings, after having been placed in circuit with a Daniell cell, and their high resistance observed, may then be completely insulated and submitted in this condition to the action of a distant spark, or of a charged rod, or of induced currents. If, after this, the filings are replaced in their original circuit, the enormous increase in their conductivity is immediately apparent.

The conductivity produced by these various methods takes place throughout the whole mass of the metallic filings, and in every direction, as the following experiment will show. A vertical ebonite cup containing aluminium powder (fig. 2) is placed between two metal plates A, B; laterally the powder is in contact with two short rods C, D, which pass through the sides of the ebonite cylinder. A and B can be connected to two terminals of one of the arms of the Wheatstone bridge, C and D being free, and *vice versâ*. Whatever arrangement is adopted, if a battery of 100 cells is joined up for a few seconds with one or the other of the pairs of terminals, the increase in the conductivity is immediately visible in that direction, and is found to exist also in the direction at right angles.

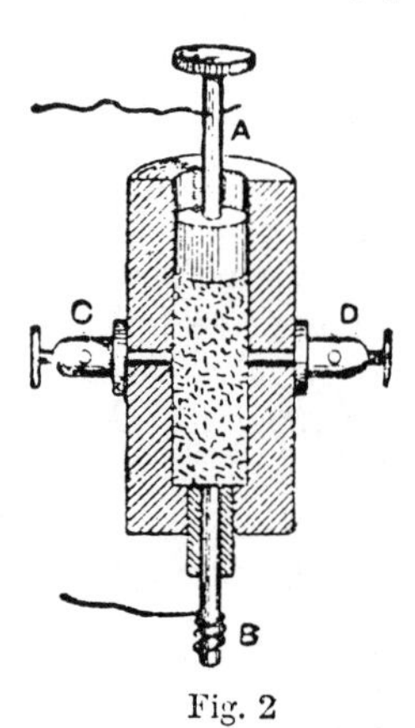

Fig. 2

Substances in which Diminution of Resistance has been observed.—The substances in which the phenomenon of the sudden increase of conductivity is most easily observed are filings of iron, aluminium, copper, brass, antimony, tellurium, cadmium, zinc, bismuth, &c. The size of the grains and their nature are not the only elements to be considered, for grains of lead of the same size, but coming from different quarters, offer at the same temperature great differences in resistance (20,000 to 500,000 ohms). Extremely fine metallic powder, as a rule, offers almost perfect resistance to the passage of a current. But if we take a sufficiently short column and exert a sufficiently great pressure, a point is soon reached when the electrical influence will effect a sudden increase in the conductivity.

Thus, a layer of copper reduced by hydrogen, which does not become conducting under the influence of the electric spark or otherwise, will become so on being submitted to a pressure of 500 grammes to the square centimetre (7 lb. per square inch). Instead of using pressure, I employed as a conductor in some experiments a very fine coating of powdered copper spread on a sheet of unpolished glass or ebonite E (fig. 3), 7 centimetres long and 2 centimetres broad. A layer of this kind, polished with a burnisher, has a very variable resistance. With a little care one can prepare sheets which are more or less sensitive to electrical action.

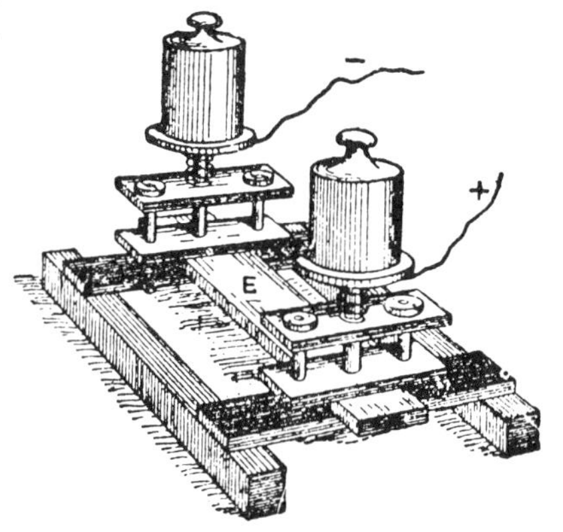

Fig. 3.

Metal powders or metal filings are not the only sensitive substances, as powdered galena, which is slightly conducting under pressure, conducts much better after having been submitted to electrical influence. Powdered binoxide of manganese is not very sensitive unless mixed with powdered antimony and compressed.

Making use of the A arrangement with very short sparks at S (fig. 1), the phenomenon of increased conductivity can be observed with platinised and silvered glass, also with glass covered with gold, silver, and aluminium foil. Some of the mixtures employed had the consistency of paste. These were mixtures of colza oil and iron, or antimony filings, and of ether or petroleum and aluminium, and plumbago, &c. Other mixtures were solid. If we make a mixture of iron filings and Canada balsam, melted in a water bath, and pour the paste into a little ebonite cup, the ends of which are closed by metallic rods, a substance is obtained which solidifies on cooling. The resistance of such a mixture is lowered from several megohms to a few hundred ohms by an electric spark. Similar results are obtained with a solid rod composed of fused flowers of sulphur and iron or aluminium filings, also by a mixture of melted resin and aluminium filings. In the preparation of these solid sensitive mixtures, care must be taken that the insulating substance should only form a small percentage of the whole.

Some interesting results are also obtained with mixtures of sulphur and aluminium, and with resin and aluminium, when in a state of powder. When cold these mixtures, as a rule, do not conduct either directly or after they have been exposed to electrical influences, but they become conducting on combining pressure with electrical influences. Thus, a mixture of flowers-of-sulphur and aluminium filings in equal volumes was placed in a glass tube 24 mm. in diameter. The weight of the mixture was 20 grammes, and the height of the column 22 mm., with a pressure of 186 grammes per square centimetre (2½ lb. per square inch). The mixture is not conducting, but after exposure to electrical influence, obtained by the A arrangement, the resistance falls to 90 ohms. In a similar manner a mixture of selenium and aluminium, placed in a tube 99 mm. long, was not conducting until after it was exposed to the combined influence of pressure and electricity.

The following is one of the group of numerous experiments of a slightly different character. A mixture of flowers-of-sulphur and fine aluminium filings, containing two of sulphur to one of aluminium, is placed in a cylindrical glass tube 35 mm. long. By means of a piston, a pressure of 20 kilogrammes per square centimetre (284 lb. per square inch) was applied. It was only necessary to connect the column for 10 seconds to the poles of a 25-cell battery, for the resistance originally infinite to be reduced to 4000 ohms.

The arrangement shown in fig. 4 illustrates another order of experiment. Two rods of copper were oxidised in the flame of a Bunsen burner, and were then arranged to lie across each other, as shown, and were connected to the terminals of the arm of a Wheatstone bridge, the high resistance of the circuit being due to the layers of oxide. Amongst the many measurements made, I found, in one case, a resistance of 80,000 ohms, which, after exposure to the influence of the electric spark, was reduced to 7 ohms. Analogous effects are obtained with oxidised steel rods. Another pretty experiment is to place a cylinder of copper, with an oxidised hemispherical head, on a sheet of oxidised copper. Before exposure to the influence of the electric spark, the oxide offers considerable resistance. The experiment can be repeated several times by merely moving the cylinder from one place to another on the oxidised sheet of copper, thus showing that the phenomenon only takes place at the point of contact of the two layers of oxide.

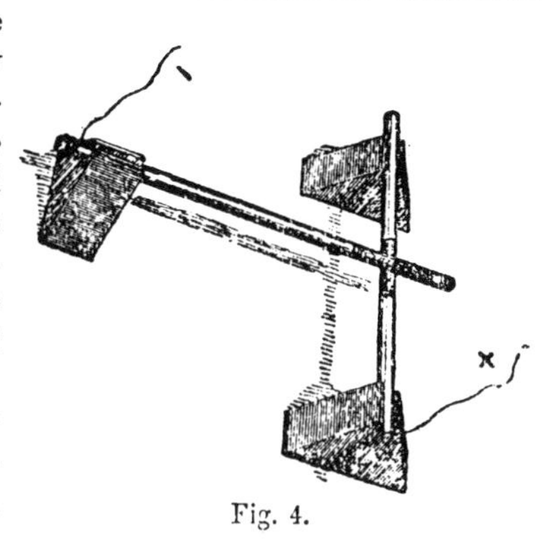

Fig. 4.

It may be worth noting that, for most of the substances enumerated, an elevation of temperature diminishes the resistance, but the effect of a rise of temperature is transient, and is incomparably less than the effect due to currents of high potential. For a few substances the two effects are opposed.

Restoration of Original Resistance.—The conductivity caused by the various electrical influences lasts sometimes for a long period (twenty-four hours or more), but it is always possible to make it rapidly disappear, particularly by a shock.

The majority of substances tested showed an increase of resistance on being shaken previous to being submitted to any special electrical influence, but after having been influenced the effect of shock is much more marked. The phenomenon is best seen with the metallic filings, but it can also be observed with metallised ebonite sheets with mixtures of liquid insulators and metallic powders, mixtures of metallic filings and insulating substances (compressed or not compressed), and finally with solid bodies.

I observed the return to original resistance in the following manner :—

The sensitive substance was placed at K (fig. 1), and formed part of a circuit which included a Daniell cell and galvanometer. At first no current passes. Sparks are then caused at S, and the needle of the galvanometer is permanently deflected. On smartly tapping the table supporting the ebonite cup in which the sensitive substance is contained, the original condition is completely restored. When the electric action has been of a powerful character, violent blows are necessary. I employed for the purpose of these shocks a hammer fixed on the table, the blows of which could be regulated.

With some substances, when feebly electrified, the return seemed to be spontaneous, although it was slower than the return of the galvanometer needle to equilibrium. This restoration of the original resistance is attributable to surrounding trepidations, as it was only necessary to walk about the room at a distance of a few metres, or to shake a distant wall. This spontaneous return to original resistance after weak electrical action was visible with a mixture of equal parts of fine selenium and tellurium powders. The restoration of resistance by shock was not observable so long as the electrical influence was at work.

After having been submitted to powerful electric action, shock does not seem to entirely restore substances to their original state; in fact, the substances generally show greater sensitiveness to electric action. Thus, a mixture of colza oil and antimony powder being exposed to the influence of arrangement A, a spark of 5 mm. was at first necessary to break down the resistance; but after the conductivity had been made to disappear by means of blows, a spark of only 1 mm. was sufficient to again render the substance conducting. Finely powdered aluminium has an extremely high resistance. A vertical column of powdered aluminium 5 mm. long of 4 square cms. cross-section, submitted to considerable pressure, completely stopped the current from a Daniell cell. The influence of arrangement A produced no effect, but, by direct contact with a Leyden jar, the resistance was reduced to 50 ohms. The effect of shock was then tried, and after this the sparks produced by arrangement A were able to reduce the resistance.

The following experiment is also of the same kind: Aluminium filings placed in a parallelipidic trough completely stopped the current from a Daniell cell, and the resistance offered to a single cell remained infinite after the trough had been placed in the circuit of 25 sulphate of mercury cells for 10 seconds. The aluminium was next placed in circuit with a battery of 75 cells ; a single Daniell cell was then able to send a current through the substance. The original resistance was restored by shock, but not the original condition of things, since a single cell was able to send a current after the aluminium had been circuited for 10 seconds with a battery of only 25 cells. I may add that if the restoration of resistance was brought about by a *violent* shock, it was necessary to place the aluminium in circuit with 75 cells for one minute before the resistance was again broken down.

It must be observed that electrical influence is not always necessary to restore conductivity after an apparent return to the original resistance, repeated feeble blows being sometimes successful in bringing this about. Both in the case of slow return by time and sudden return by shock, the original value of the resistance is often increased. Rods of Carré carbon, 1 metre long and 1 mm. in diameter, were particularly noticeable for this phenomenon.

Return to Original Resistance by Temperature Elevation.—A plate of coppered ebonite rendered conducting by electricity, and placed close to a gas-jet, quickly regained its original resistance. A solid rod of resin and aluminium, or of sulphur and aluminium, rendered conducting by connection to the poles of a small battery, will regain its original resistance by shock ; but if the conducting state has been caused by powerful means, such, for instance, as direct contact with a Leyden jar, shock no longer has any effect, at least such a shock as the fragile nature of the material can stand. A slight rise of temperature, however, has the desired result. By suitably regulating the electric action it is possible to get a substance into such a condition that the warmth of the fingers suffices to annul conductivity.

Influence of Surroundings.—Electric action gives rise to no alteration of resistance when the substance is entirely within a closed metal box. The sensitive substance, in circuit with a Daniell cell and a galvanometer, is placed inside a brass box (fig. 5). The absence of current is ascertained, the circuit broken, and the box closed. A Wimshurst machine is then worked a little way off, and will be found to have had no effect. The same result will be obtained if the circuit is kept closed during the time the Wimshurst machine is in operation.

If a wire connected at some point to the circuit is passed out through a hole in the box to a distance of 20 to 50 cms., the influence of the Wimshurst machine makes itself felt. On tapping the lid to restore resistance, the galvanometer needle remains deflected so long as the sparks continue to pass. If, however, the wires are pushed in so that they only project a few millimetres, the sparks still passing, a few taps suffice to bring back the needle to zero. On touching the end of the wire with the fingers or a piece of metal, conductivity is immediately restored. The movements of the galvanometer needle were rendered visible in these experiments by looking through a piece of wide-mesh wire-gauze with a telescope. The respective position of the things was reversed; that is to say, a Ruhmkorff coil and a periodically discharged Leyden jar were placed inside, and the sensitive substance outside, the box, with the same results.

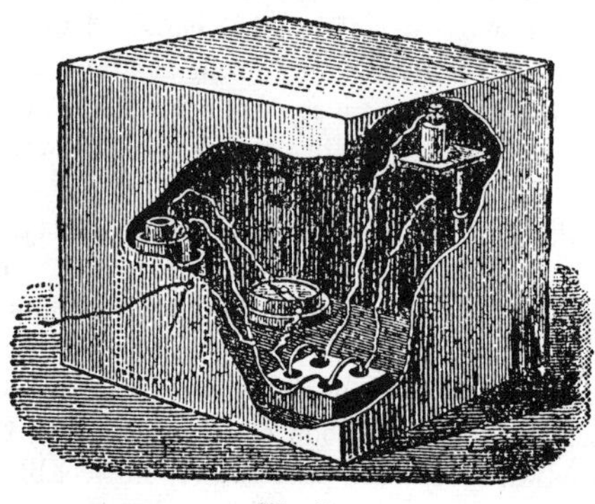

Fig. 5.

In some later experiments with a larger metallic case, and with the Daniell cell, sensitive substance, delicate galvanometer, and Wheatstone bridge placed inside, I found that a double casing was necessary in order to absolutely suppress all effects. A glass covering afforded no protection.

Considerations on the Mechanism of the Effects produced.—What conclusions are we to draw from the experiments described? The substances employed in these investigations were not conductors, since the metallic particles composing them were separated from each other in the midst of an insulating medium. It was not surprising that currents of high potential, and especially currents induced by discharges, should spark across the insulating intervals. But as the conductivity *persisted* afterwards, even for the weakest thermo electric currents, there is some ground for supposing that the insulating medium is transformed by the passage of the current, and that certain actions, such as shock and rise of temperature, bring about a modification of this new state of the insulating body. Actual movement of the metallic particles cannot be imagined in experiments where the particles in a layer a few millimetres thick were fixed in an invariable relative position by extreme pressures, reaching at times to more than 100 kilogrammes per square cm. (1420 lb. to the square inch). Moreover, in the case of solid mixtures, in which the same variations of resistance were produced, displacement seems out of the question. To explain the persistence of the conductivity after the cessation of the electrical influence, are we to suppose in the case of metallic filings a partial volatilisation of the particles creating a conducting medium between the grains of metal? In the case of mixtures of metallic powders and insulating substances agglomerated by fusion, are we to suppose that the thin insulating layers are pierced by the passage of very small sparks, and that the holes left behind are coated with conducting material? If this explication is admissible for induced currents, it must hold good for continuous currents. If so, we must conclude that these mechanical actions may be produced by batteries of only 10 to 20 volts electromotive force, and which only cause an insignificant current to pass. The following experiment is worth quoting in this connection:—

A circuit was formed by a Daniell cell, a sensitive galvanometer, and some aluminium filings in an ebonite cup. The galvanometer needle remained at zero. The filings were cut out of this circuit, and switched for one minute into circuit with a battery of 43 sulphate of mercury cells. On being replaced in the first circuit, the filings exhibited high conductivity. The result was the same when 10 or 20 cells were employed, or when the current was diminished by interposing in the circuit a column of distilled water, 40 cm. long and 20 mm. in diameter. The cells used (platinum, sulphate of mercury, sulphate of zinc, zinc) had a high internal resistance. Thus, 43 cells (60 volts), when short circuited, only gave a current of 5 milliampères. The same battery, with the column of distilled water in circuit only, caused a deflection of 100 mm. on a scale one metre off, with an astatic galvanometer wound with 50,000 turns. We can, therefore, see how infinitesimally small the initial current must have been when the filings were added to the circuit. The battery acted, therefore, essentially by virtue of its electromotive force.

If mechanical displacement of particles or transportation of conducting bodies seem inadmissible, it is probable that there is a modification of the insulator itself, the modification persisting for some time by virtue of a sort of "coercive force." An electric current of high potential, which would be completely stopped by a thick insulating sheet, may be supposed to gradually traverse the very thin dielectric layers between the conducting particles, the passage being effected very rapidly if the electric pressure is great, and more slowly if the pressure is less.

Increase of Resistance.—An increase of resistance was observed in these investigations less often than a diminution ; nevertheless, a number of frequently repeated experiments enable me to say that increase of resistance is not exceptional, and that the conditions under which it takes place are well defined. Short columns of antimony or aluminium powder, when subjected to a pressure of about 1 kilogramme per square cm. (14·2 lb. per square inch), and offering but a low resistance, exhibited an increase of resistance under the influence of a powerful electrification. Peroxide of lead, a fairly good conductor, always exhibited an increase, so also did some kinds of platinised glass, while others showed alternate effects. For instance, a sheet of platinised glass, which offered a resistance of 700 ohms, became highly conducting after 150 sulphate of mercury cells had been applied to it for 10 seconds. This condition of conductivity was annulled by contact with a charged Leyden jar, and reappeared after again applying 150 cells for 10 seconds, and so on. Similar effects were obtained with a thin layer of a mixture of selenium and tellurium poured, when fused, into a groove in a sheet of mica placed between two copper plates. These alternations were always observed several times in succession, and at intervals of several days.

These augmentations and alternations are in no way incompatible with the hypothesis of a physical modification of the insulator by electrical influence.

APPENDIX D.

RESEARCHES OF PROF. D. E. HUGHES, F.R.S., IN ELECTRIC WAVES AND THEIR APPLICATION TO WIRELESS TELEGRAPHY, 1879-1886.

It may be desirable to place briefly on record the circumstances under which the following remarkable communication was written.

While revising the last sheets of this work, it occurred to the author to ask Sir William Crookes for some particulars of the experiments to which he alluded in his 'Fortnightly' article, some passages from which are quoted on pp. 201-203. On April 22, 1899, Sir William replied as follows :—

DEAR MR FAHIE,—The experiments referred to at page 176 of my 'Fortnightly' article as having taken place "some years ago" were tried by Prof. Hughes when experimenting with the microphone.

I have not ceased since then urging on him to publish an account of his experiments. I do not feel justified in saying more about them, but if you were to write to him, telling him what I say, it might induce him to publish.

It is a pity that a man who was so far ahead of all other workers in the field of wireless telegraphy should lose all the credit due to his great ingenuity and prevision.—Believe me, very truly yours, WILLIAM CROOKES.

On receipt of this letter I wrote to Prof. Hughes. In reply he said :—

"Your letter of 26th instant has brought upon me a flood of old souvenirs in relation to my past experiments on aërial telegraphy. They were completely unknown to the general public, and I feared that the few distinguished men who saw them had forgotten them, or at least had forgotten how the results shown them were produced. . . .

"At this late date I do not wish to set up any claim to

priority, as I have never published a word on the subject; and it would be unfair to later workers in the same field to spring an unforeseen claimant to the experiments which they have certainly made without any knowledge of my work."

On second (and my readers will say, wiser) thoughts, Prof. Hughes sent me the following letter, in the eliciting of which I consider myself especially fortunate and privileged :—

40 Langham Street, W., *April* 29, 1899.

Dear Sir,—In reply to yours of the 26th inst., in which you say that Sir William Crookes has told you that he saw some experiments of mine on aërial telegraphy in about December 1879, of which he thinks I ought to have published an account, and of which you ask for some information, I beg to reply with a few leading experiments that I made on this subject from 1879 up to 1886 :—

In 1879, being engaged upon experiments with my microphone, together with my induction balance, I remarked that at some times I could not get a perfect balance in the induction balance, through apparent want of insulation in the coils ; but investigation showed me that the real cause was some loose contact or microphonic joint excited in some portion of the circuit. I then applied the microphone, and found that it gave a current or sound in the telephone receiver, no matter if the microphone was placed direct in the circuit or placed independently at several feet distance from the coils, through which an intermittent current was passing. After numerous experiments, I found that the effect was entirely caused by the extra current, produced in the primary coil of the induction balance.

Further researches proved that an interrupted current in any coil gave out at each interruption such intense extra currents that the whole atmosphere in the room (or in several rooms distant) would have a momentary invisible charge, which became evident if a microphonic joint was used as a receiver with a telephone. This led me to experiment upon the best form of a receiver for these invisible electric waves, which evidently permeated great distances, and through all apparent obstacles, such as walls, &c. I found that all microphonic contacts or joints were extremely sensitive. Those formed of a hard carbon such as coke, or a combination of a piece of coke resting upon a bright steel contact, were very sensitive and self-restoring; whilst a loose contact between metals was equally sensitive, but would cohere, or remain in full contact, after the passage of an electric wave.

The sensitiveness of these microphonic contacts in metals has since been rediscovered by Mons. Ed. Branly of Paris, and by Prof. Oliver Lodge, in England, by whom the name of "coherer" has been given to this organ of reception; but, as we wish this organ to make a momentary contact and not cohere permanently, the name seems to me ill-suited for the instrument. The most sensitive and perfect receiver that I have yet made does not cohere permanently, but recovers its original state instantly, and therefore requires no tapping or mechanical aid to the separation of the contacts after momentarily being brought into close union.

I soon found that, whilst an invisible spark would produce a thermo-electric current in the microphonic contacts (sufficient to be heard in the telephone in its circuit), it was far better and more powerful to use a feeble voltaic cell in the receiving circuit, the microphonic joint then acting as a relay by diminishing the resistance at the contact, under the influence of the electric wave received through the atmosphere.

I will not describe the numerous forms of the transmitter and receiver that I made in 1879, all of which I wrote down in several volumes of manuscripts in 1879 (but these have never been published), and most of which can be seen here at my residence at any time; but I will confine myself now to a few salient points. I found that very sudden electric impulses, whether given out to the atmosphere through the extra current from a coil or from a frictional electric machine, equally affected the microphonic joint, the effect depending more on the sudden high potential effect than on any prolonged action. Thus, a spark obtained by rubbing a piece of sealing-wax was equally effective as a discharge from a Leyden jar of the same potential.[1] The rubbed sealing-wax and charged Leyden jar had no effect until they were discharged by a spark, and it was evident that this spark, however feeble, acted upon the whole surrounding atmosphere in the form of waves or invisible rays,

[1] Prof. Lodge subsequently and independently observed this fact, and illustrates it beautifully in his 'Work of Hertz,' pp. 27, 28.—J. J. F.

the laws of which I could not at the time determine. Hertz, however, by a series of original and masterly experiments, proved in 1887-89 that they were real waves similar to light, but of a lower frequency, though of the same velocity. In 1879, whilst making these experiments on aërial transmission, I had two different problems to solve: 1st, What was the true nature of these electrical aërial waves, which seemed, whilst not visible, to spurn all idea of insulation, and to penetrate all space to a distance undetermined. 2nd, To discover the best receiver that could act upon a telephone or telegraph instrument, so as to be able to utilise (when required) these waves for the transmission of messages. The second problem came easy to me when I found that the microphone, which I had previously discovered in 1877-78, had alone the power of rendering these invisible waves evident, either in a telephone or a galvanometer, and up to the present time I do not know of anything approaching the sensitiveness of a microphonic joint as a receiver. Branly's tube, now used by Marconi, was described in my first paper to the Royal Society (May 8, 1878) as the microphone tube, filled with loose filings of zinc and silver; and Prof. Lodge's coherer is an ordinary steel microphone, used for a different purpose from that in which I first described it.[1]

During the long-continued experiments on this subject, between 1879 and 1886, many curious phenomena came out which would be too long to describe. I found that the effect

[1] Prof. Hughes is rightly regarded as the real discoverer of the electrical behaviour of a bad joint or loose contact, the study of which in his hands has given us the microphone; but as in the case of Hertzian-wave effects before Hertz, so, long before Hughes, "mere phenomena of loose contact," as Sir George Stokes called them, must have often manifested themselves in the working of electrical apparatus. For an interesting example see Arthur Schuster's paper read before the British Association in 1874 (or abstract, 'Telegraphic Journal,' vol. ii. p. 289), where the effects are described as a new discovery in electricity, and disguised under the title of the paper, "On Unilateral Conductivity." Schuster suspected the cause—"Two wires screwed together may not touch each other, but be separated by a thin layer of air"—but he missed its real significance. The phenomenon was a kind of bye-product, cropped up while he was engaged on other work, and so was not pursued far enough. —J. J. F.

of the extra current in a coil was not increased by having an iron core as an electro-magnet—the extra current was less rapid, and therefore less effective. A similar effect of a delay was produced by Leyden-jar discharges. The material of the contact-breaker of the primary current had also a great effect. Thus, if the current was broken between two or one piece of carbon, no effect could be perceived of aërial waves, even at short distances of a few feet. The extra current from a small coil without iron was as powerful as an intense spark from a secondary coil, and at that time my experiments seemed to be confined to the use of a single coil of my induction balance, charged by six Daniell cells. With higher battery power the extra current invariably destroyed the insulation of the coils.

In December 1879 I invited several persons to see the results then obtained. Amongst others who called on me and saw my results were—

Dec. 1879.—Mr W. H. Preece, F.R.S.; Sir William Crookes, F.R.S.; Sir W. Roberts-Austen, F.R.S.; Prof. W. Grylls Adams, F.R.S.; Mr W. Grove.

Feb. 20, 1880.—Mr Spottiswoode, Pres. R.S.; Prof. Huxley, F.R.S.; Sir George Gabriel Stokes, F.R.S.

Nov. 7, 1888.—Prof. Dewar, F.R.S.; Mr Lennox, Royal Institution.

They all saw experiments upon aërial transmission, as already described, by means of the extra current produced from a small coil and received upon a semi-metallic microphone, the results being heard upon a telephone in connection with the receiving microphone. The transmitter and receiver were in different rooms, about 60 feet apart. After trying successfully all distances allowed in my residence in Portland Street, my usual method was to put the transmitter in operation and walk up and down Great Portland Street with the receiver in my hand, with the telephone to the ear.

The sounds seemed to slightly increase for a distance of 60 yards, then gradually diminish, until at 500 yards I could no longer with certainty hear the transmitted signals. What struck me as remarkable was that, opposite certain houses, I could hear better, whilst at others the signals could hardly be perceived. Hertz's discovery of nodal points in reflected waves (in 1887-89) has explained to me what was then considered a mystery.

At Mr A. Stroh's telegraph instrument manufactory Mr Stroh and myself could hear perfectly the currents transmitted from the third storey to the basement, but I could not detect clear signals at my residence about a mile distant. The innumerable gas and water pipes intervening seemed to absorb or weaken too much the feeble transmitted extra currents from a small coil.

The President of the Royal Society, Mr Spottiswoode, together with the two hon. secretaries, Prof. Huxley and Prof. G. Stokes, called upon me on February 20, 1880, to see my experiments upon aërial transmission of signals. The experiments shown were most successful, and at first they seemed astonished at the results ; but towards the close of three hours' experiments Prof. Stokes said that all the results could be explained by known electro-magnetic induction effects, and therefore he could not accept my view of actual aërial electric waves unknown up to that time, but thought I had quite enough original matter to form a paper on the subject to be read at the Royal Society.

I was so discouraged at being unable to convince them of the truth of these aërial electric waves that I actually refused to write a paper on the subject until I was better prepared to demonstrate the existence of these waves ; and I continued my experiments for some years, in hopes of arriving at a perfect scientific demonstration of the existence of aërial electric waves produced by a spark from the extra currents in coils, or from frictional electricity, or from secondary coils. The triumphant demonstration of these waves was reserved to Prof. Hertz, who by his masterly researches upon the subject in 1887-89 completely demonstrated not only their existence but their identity with ordinary light, in having the power of being reflected and refracted, &c., with nodal points, by means of which the length of the waves could be measured. Hertz's experiments were far more conclusive than mine, although he used a much less effective receiver than the microphone or coherer.

I then felt it was now too late to bring forward my previous experiments ; and through not publishing my results and means employed, I have been forced to see others remake the discoveries I had previously made as to the sensitiveness of the microphonic contact and its useful employment as a receiver for electric aërial waves.

Amongst the earliest workers in the field of aërial transmission I would draw attention to the experiments of Prof. Henry, who describes in his work, published by the Smithsonian Institute, Washington, D.C., U.S.A., vol. i. p. 203 (date unknown, probably about 1850), how he magnetised a needle in a coil at 30 feet distance, and magnetised a needle by a discharge of lightning at eight miles' distance.[1]

Marconi has lately demonstrated that by the use of the Hertzian waves and Branly's coherer he has been enabled to transmit and receive aërial electric waves to a greater distance than previously ever dreamed of by the numerous discoverers and inventors who have worked silently in this field. His efforts at demonstration merit the success he has received ; and if (as I have lately read) he has discovered the means of concentrating these waves on a single desired point without diminishing their power, then the world will be right in placing his name on the highest pinnacle in relation to aërial electric telegraphy.—Sincerely yours, D. E. HUGHES.

J. J. FAHIE, Esq.,
Claremont Hill, St Helier's, Jersey.

On the publication of this letter in the 'Electrician' (May 5, 1899), Mr John Munro called on Prof. Hughes, and was accorded the privilege of inspecting his apparatus, mostly self-made and of the simplest materials, and his note-books, filled with experiments in ink or pencil, dated or dateless, and some marked "extraordinary," "important," and so on. An interest-

[1] The 'Polytechnic Review,' March 25, 1843, says: "Professor Henry communicated to the American Society that he had succeeded in magnetising needles by the secondary current in a wire more than 220 feet distant from the wire through which the primary current, excited by a single spark from an electrical machine, was passing." Indeed Prof. Henry noted many cases of what we now call Hertzian-wave effects, but what he and every one else in those days thought were only extraordinary cases of induction. Many experimenters after Henry must have observed similar effects. See for example 'Telegraphic Journal,' February 15, 1876, p. 61, on "The 'Etheric' Force"; and the 'Electrician,' vol. xliii. p. 204.—J. J. F.

ing account of this interview was afterwards published by Mr Munro,[1] from which I make a few extracts, as they help to illustrate and supplement the Professor's own account.

After satisfying himself as to the cause of the trouble in his induction-balance experiments as stated above (p. 296), Prof. Hughes joined a single cell B (fig. 1) in circuit with a clockwork interrupter I, and the primary coil C of the induction balance. This "transmitter" was connected by a wire W, several feet in length, to the "receiver," which consisted of a telephone T in circuit with a microphone M. With such an arrangement the "extra spark" of the transmitter was always heard in the telephone. These sounds were found to vary with the conditions of the experiment: thus, with an electromotive force of $\frac{1}{50}$ volt the sound was stronger than with several cells; it was also louder and clearer when the contact points of the

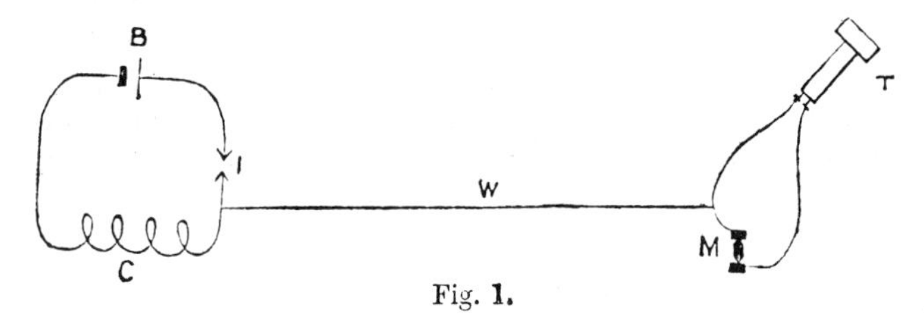

Fig. 1.

interrupter were of metal—not metal to carbon, or carbon to carbon. Again, an iron core in the coil C, though productive of a stronger spark, rather diminished than increased the corresponding sound in the telephone. Indeed, the spark from the Faraday electro-magnet of the Royal Institution, excited by a large Grove battery, had little effect, and even a dynamo at work beside the receiver gave a very poor result.

Prof. Hughes tried many experiments to satisfy himself that his receiver (his microphone and telephone) was influenced by the extra spark solely, and not by the ordinary electro-magnetic induction. He inserted coils in the transmitting and receiving circuits, placing them parallel, and at right angles to each other —that is, in positions favourable and unfavourable to such induction—but without modifying the effect. He also reduced the number of turns of wire on the coil C, and even removed

[1] 'Electrical Review,' June 2, 1899.

it altogether, connecting the battery and interrupter by only three inches of wire, and still heard the sounds as distinctly as before. That electro-static induction had no part in the phenomenon was shown by inserting charged conductors of large surface (for example metal discs) in the two circuits and shifting their positions with respect to each other without producing any effect on the receiver.

Having concluded from these and numerous other observations that the results were conductive in principle rather than inductive, and were due to electrical impulses or waves set in motion by the sparks at the interrupter and filling all the surrounding space, Prof. Hughes set himself to find the most sen-

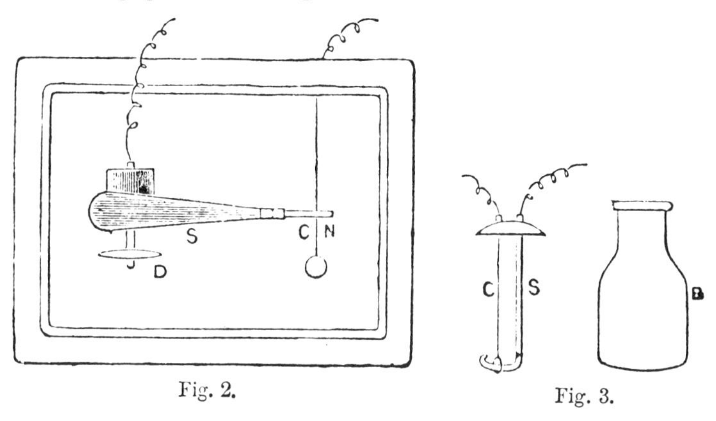

Fig. 2. Fig. 3.

sitive form of microphone to receive the waves. Contacts of metal were found to be apt to stick together, or "cohere," as we now say. A microphone which is both sensitive and self-restoring or non-cohering is made with a carbon contact resting lightly on bright steel, as shown in fig. 2, where C is a carbon pencil touching a needle N, and S an adjustable spring of brass by which the pressure of the contact can be regulated by means of the disc D. An extremely sensitive but easily deranged form of microphone is shown in fig. 3, where S is a steel hook, and C a fine copper wire with a loop on the end which has been oxidised and smoked in the flame of a spirit-lamp. The carbonised loop and steel hook are placed in a small bottle B for safety.

Another form of microphone which the Professor tried was a tube containing metal filings, which forestalls the Branly tube, but as the coherence of the filings was a disadvantage he abandoned it. Contacts of iron and mercury were sensitive, but very troublesome ; while contacts of iron and steel cohered, but were sensitive, and kept well when immersed in a mixture of petroleum and vaseline, which, though an insulator, does not bar the electric waves.

Some of these microphone arrangements were found to be very sensitive to small charges of electricity—far more so than the gold-leaf electroscope and the quadrant electrometer. Even a metal filing on a stick of sealing-wax carried enough electricity from a Leyden jar to affect the microphone and give a

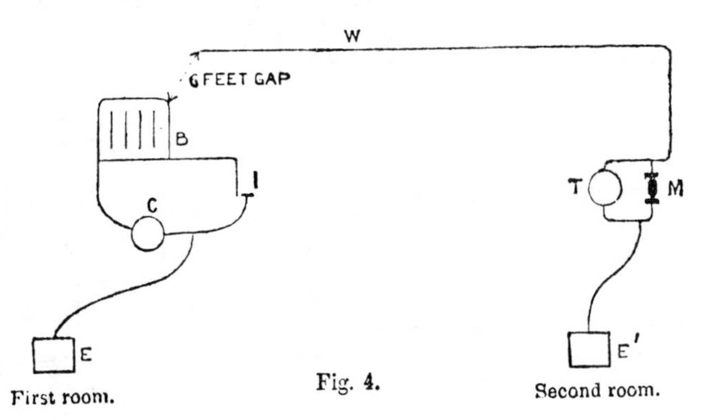

Fig. 4.

sound in the telephone, while it had no effect on the electroscope or the electrometer.

With such delicate receivers Prof. Hughes discarded the connecting wire W in fig. 1, thus separating the receiver from the transmitter, and producing the germ of a wireless telegraph. His first experiment of this kind was made between October 15 and 24, 1879, the transmitter being in one room and the receiver in an adjoining room, but a wire from the receiver limited the air gap to about 6 feet. Fig. 4, which is roughly copied from the Professor's own diagram, shows the arrangement, where W is the wire, B the battery, I the interrupter, C the coil, T the telephone, M the microphone, and E, E' the earth (gas-pipes). In another experiment, made about the middle of November 1879, he connected a fender to the interrupter "to act as a radiator," and afterwards, instead of the fender, he used wires (answering to the "wings" of Hertz) on both transmitting and receiving apparatus, the wires being stiffened with laths to hold them in place.

The use of an "earth" connection led him to try the effect of joining the telephone to a gas-pipe of lead, and the microphone to a water-pipe of iron, as shown in fig. 5. The result was an improved sound in the telephone, and he concluded that the different metals formed a weak "earth battery," from which a permanent current ran through the circuit. On this supposition he reasoned that the electric waves influencing the microphone, and perhaps changing its resistance, would rapidly alter the strength of this current, and so account for the heightened effects in the telephone. Acting on this idea, he included an E.M.F. in the receiving circuit. A single cell was more than enough, and had to be reduced to as little as $\frac{1}{25}$th of a volt in order not to permanently break down the contact resistance of the microphone.

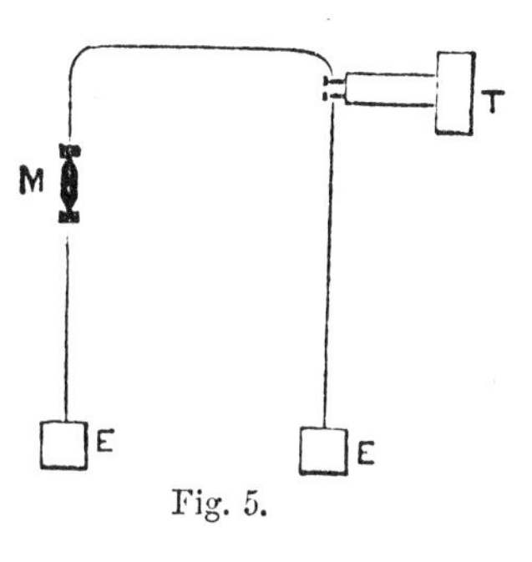

Fig. 5.

"Thus," says Mr Munro, "Prof. Hughes had step by step put together all the principal elements of the wireless telegraph as we know it to-day, and although he was groping in the dark before the light of Hertz arose, it is little short of magical that in a few months, even weeks, and by using the simplest means, he thus forestalled the great Marconi advance by nearly twenty years!"

In the fifty years (just completed) of a brilliant professorial career at Cambridge, Sir George Stokes has given, times out of number, sound advice and helpful suggestions to those who have sought him; but in this case, as events show, the great weight of his opinion has kept back the clock for many years. With proper encouragement in 1879-80 Prof. Hughes would have followed up his clues, and, with his extraordinary keenness in research, there can be no doubt that he would have anticipated Hertz in the complete discovery of electric waves, and Marconi in the application of them to wireless telegraphy, and so have altered considerably the course of scientific history.

As a recent commentator pithily says: "Hughes's experiments of 1879 were virtually a discovery of Hertzian waves before Hertz, of the coherer before Branly, and of wireless telegraphy before Marconi and others." The writer goes on to say, "Prof. Hughes has a great reputation already, but these latter experiments will add enormously to it, and place him among the foremost electricians of all time"[1]—praise which, knowing the learned professor as I do, I consider none too great.

APPENDIX E.

Reprint of Signor G. Marconi's Patent.

No. 12,039, A.D. 1896.

Date of Application, 2*nd June* 1896. *Complete Specification Left*, 2*nd Mar.* 1897; *Accepted*, 2*nd July* 1897.

PROVISIONAL SPECIFICATION.

IMPROVEMENTS IN TRANSMITTING ELECTRICAL IMPULSES AND SIGNALS, AND IN APPARATUS THEREFOR.

I, Guglielmo Marconi, of 71 Hereford Road, Bayswater, in the county of Middlesex, do hereby declare the nature of this invention to be as follows:—

According to this invention electrical actions or manifestations are transmitted through the air, earth, or water by means of electric oscillations of high frequency.

At the transmitting station I employ a Ruhmkorff coil having in its primary circuit a Morse key, or other appliance

[1] The 'Globe,' May 12, 1899. Prof. Hughes died, full of honours, on January 22, 1900, aged sixty-nine. See, amongst other obituary notices, the 'Times,' January 24, and the 'Electrician,' January 26.

for starting or interrupting the current, and its pole appliances (such as insulated balls separated by small air spaces or high vacuum spaces, or compressed air or gas, or insulating liquids kept in place by a suitable insulating material, or tubes separated by similar spaces and carrying sliding discs) for producing the desired oscillations.

I find that a Ruhmkorff coil, or other similar apparatus, works much better if one of its vibrating contacts or brakes on its primary circuit is caused to revolve, which causes the secondary discharge to be more powerful and more regular, and keeps the platinum contacts of the vibrator cleaner and preserves them in good working order for an incomparably longer time than if they were not revolved. I cause them to revolve by means of a small electric motor actuated by the current which works the coil, or by another current, or in some cases I employ a mechanical (non-electrical) motor.

The coil may, however, be replaced by any other source of high tension electricity.

At the receiving instrument there is a local battery circuit containing an ordinary receiving telegraphic or signalling instrument, or other apparatus which may be necessary to work from a distance, and an appliance for closing the circuit, the latter being actuated by the oscillations from the transmitting instrument.

The appliance I employ consists of a tube containing conductive powder, or grains, or conductors in imperfect contact, each end of the column of powder or the terminals of the imperfect contact or conductor being connected to a metallic plate, preferably of suitable length so as to cause the system to resonate electrically in unison with the electrical oscillations transmitted to it. In some cases I give these plates or conductors the shape of an ordinary Hertz resonator consisting of two semicircular conductors, but with the difference that at the spark-gap I place one of my sensitive tubes, whilst the other ends of the conductors are connected to small condensers.

I have found that the best rules for making the sensitive tubes are as follows :—

1st. The column of powder ought not to be long, the effects being better in sensitiveness and regularity with tubes containing columns of powder or grains not exceeding two-thirds of an inch in length.

2nd. The tube containing the powder ought to be sealed.

3rd. Each wire which passes through the tube, in order to establish electrical communication, ought to terminate with pieces of metal or small knobs of a comparatively large surface, or preferably with pieces of thicker wire, of a diameter equal to the internal diameter of the tube, so as to oblige the powder or grains to be corked in between.

4th. If it is necessary to employ a local battery of higher E.M.F. than that with which an ordinarily prepared tube will work, the column of powder must be longer and divided into several sections by metallic divisions, the amount of powder or grains in each section being practically in the same condition as in a tube containing a single section. When no oscillations are sent from the transmitting instrument the powder or imperfect contact does not conduct the current, and the local battery circuit is broken; but when the powder or imperfect contact is influenced by the electrical oscillations, it conducts and closes the circuit.

I find, however, that once started, the powder or contact continues to conduct even when the oscillations at the transmitting station have ceased; but if it be shaken or tapped, the circuit is broken.

I do this tapping automatically, employing the current which the sensitive tube or contact had allowed to begin to flow under the influence of the electric oscillations from the transmitting instrument to work a trembler (similar to that of an electric bell), which hits the tube or imperfect contact, and so stops the current and, consequently, its own movement, which had been generated by the said current, which by this means automatically and almost instantaneously interrupts itself until another oscillation from the transmitting instrument repeats the process. Whilst for certain purposes I prefer working the trembler and the instruments on the same circuit which contains the sensitive tube or contact, for other purposes I prefer working the trembler and the instruments on another circuit, which is made to work in accordance with the first by means of a relay. It is by means of actions from the current, which the sensitive tube or contact allows to pass when the oscillations influence it, that I prefer starting the apparatus that has to interrupt automatically the same current.

In order to prevent the action of the self-induction of the local circuits on the sensitive tube or contact, and also to destroy the perturbating effect of the small spark which occurs at the breaking of the circuit inside the tube or imperfect contact, and also at the vibrating contact of the trembler or at the movable contact of the relay, I put in derivation across those parts where the circuit is periodically broken a condenser of suitable capacity, or a coil of suitable resistance and self-induction, so that its self-induction may neutralise the self-induction of the said circuits; or preferably I employ in derivation on different parts of the circuit conductors or so-called semi-conductors of high resistance and small self-induction, such as bars of charcoal or preferably tubes containing water or other suitable liquid, in electrical communication with those conductors of the local circuits which are liable in course of self-induction to assume such differences of potential as to transmit jerky currents such as would influence the sensitive tube or contact so as to prevent its working with regularity.

In some cases, however, I find it suitable to employ an independent trembler moved by the current from another battery. This trembler is prevented from generating jerking or vibrating currents by means of the appliances which I have described. This trembler is kept going all the time during which one expects oscillations to be transmitted, and, as already described, the powder or imperfect contact closes the circuit of a local battery, in which are included the instruments which one desires to work, for the time during which the electrical oscillations are transmitted, breaking the circuit in case of the mechanical vibrations as soon as the oscillations from the transmitting machine cease. When transmitting through the air, and it is desired that the signal or electrical action should only be sent in one direction, or when it is necessary to transmit electrical effects to the greatest possible distance without wires, I place the oscillation producer at the focus or focal line of a reflector directed to the receiving station, and I place the tube or imperfect contact at the receiving instrument in a similar reflector directed towards the transmitting instrument.

When transmitting through the earth or water I connect one end of the tube or contact to earth and the other end to conductors or plates, preferably similar to each other, in the air and insulated from earth.

I find it also better to connect the tube or imperfect contact to the local circuit by means of thin wires or across two small coils of thin and insulated wire preferably containing an iron nucleus.

Dated this second day of June 1896.

GUGLIELMO MARCONI.

COMPLETE SPECIFICATION.

IMPROVEMENTS IN TRANSMITTING ELECTRICAL IMPULSES AND SIGNALS, AND IN APPARATUS THEREFOR.

I, Guglielmo Marconi, of 67 Talbot Road, Westbourne Park, formerly residing at 71 Hereford Road, Bayswater, in the county of Middlesex, do hereby declare the nature of this invention and in what manner the same is to be performed to be particularly described and ascertained in and by the following statement:—

My invention relates to the transmission of signals by means of electrical oscillations of high frequency, which are set up in space or in conductors.

In order that my specification may be understood, and before going into details, I will describe the simplest form of my invention by reference to figure 1.

In this diagram A is the transmitting instrument and B is the receiving instrument, placed at say ¼ mile apart.

In the transmitting instrument R is an ordinary induction coil (a Ruhmkorff coil or transformer).

Its primary circuit C is connected through a key D to a battery E, and the extremities of its secondary circuit F are connected to two insulated spheres or conductors G H fixed at a small distance apart.

When the current from the battery E is allowed to pass through the primary of the induction coil, sparks will take place between the spheres G H, and the space all around the spheres suffers a perturbation in consequence of these electrical rays or surgings.

The arrangement A is commonly called a Hertz radiator, and the effects which propagate through space Hertzian rays.

The receiving instrument B consists of a battery circuit J, which includes a battery or cell K, a receiving instrument L, and a tube T containing metallic powder or filings, each end

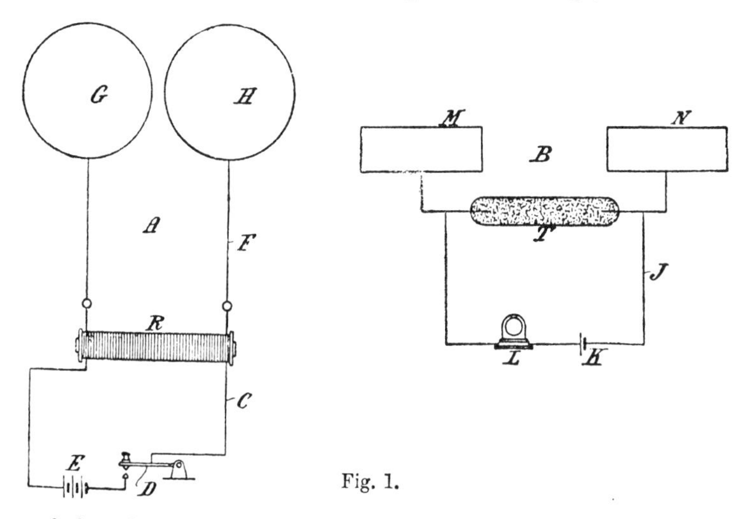

Fig. 1.

of the column of filings being also connected to plates or conductors M N of suitable size, so as to be preferably tuned with the length of wave of the radiation emitted from the transmitting instruments.

The tube containing the filings may be replaced by an imperfect electrical contact, such as two unpolished pieces of metal in light contact, or coherer, &c.

The powder in the tube T is, under ordinary conditions, a non-conductor of electricity, and the current of the cell K cannot pass through the instrument; but when the receiver is influenced by suitable electrical waves or radiation the powder in the tube T becomes a conductor (and remains so until the tube is shaken or tapped), and the current passes through the instrument.

By these means electrical waves which are set up in the

transmitting apparatus affect the receiving instrument in such a manner that currents are caused to circulate in the circuit J, and may be utilised for deflecting a needle, which thus responds to the impulse coming from the transmitter.

Figures 2, 3, 4, &c., show various more complete arrangements of the simple form of apparatus illustrated in figure 1.

I will describe these figures generally before proceeding to describe the improvements in detail.

Figure 2 is a diagrammatic front elevation of the instruments of the receiving station, in which *k k* are the plates corresponding to M N in figure 1. *g* is the battery corresponding to K, *h* is the reading instrument corresponding to L, *n* is a relay working the reading instrument *h* in the ordinary manner. *p* is a trembler or tapper, similar to that of an electric bell, which is moved by the current that works the instrument.

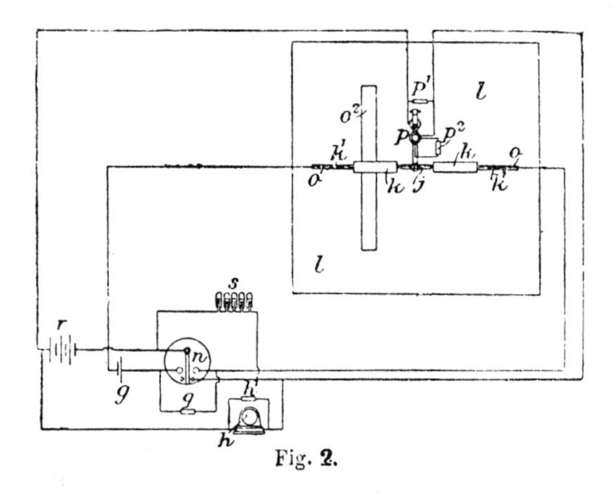

Fig. 2.

Figure 3 is a diagrammatic front elevation of the instruments at the transmitting station, in which *e e* are two metallic spheres corresponding to G H in figure 1.

c is an induction coil corresponding to R. *b* is a key corresponding to D, and *a* is a battery corresponding to E.

Figure 4 is a vertical section of the radiator or oscillation producer mounted in the focal line of a cylindrical parabolic reflector *f* in which a side view of the spheres *e e* of figure 3 is given.

Figure 5 is a full-sized view of the receiving plates *k k* and sensitive tube *j*.

Figure 5A is a modified form of sensitive tube.

Figure 6 is a modification of the oscillation producer in which the spheres *e e* and *d d* are mounted in an ebonite tube d^3.

Figure 7 is another modification of the oscillation producer in which the spheres are substituted by hemispheres.

Figure 8 is a modified form of receiver in which the plates *k k* are curved instead of being straight.

Figure 9 is another form of transmitter in which two large metallic plates t^2 t^2 are employed.

Figure 10 shows a modification of the arrangements at the transmitting station, and figure 11 a modification of the arrangements of the receiving station, which enables one to signal through obstacles such as hills or mountains.

Figure 12 shows a detector which is useful for determining the proper length of the plates *k k* of the receivers.

Figure 13 shows an improved interrupter (make-and-break) which is applicable to the induction coil of the transmitter.

Figure 14 shows a water resistance, the use of which shall be explained.

My invention relates in great measure to the manner in which the above apparatus is made and connected together. With some of these forms I am able to obtain Morse signals, and to work ordinary telegraphic instruments and other apparatus; and with modifications of the above apparatus it is possible to transmit signals not only through comparatively small obstacles such as brick walls, trees, &c., but also through or across masses of metal, or hills, or mountains, which may intervene between the transmitting and receiving instruments.

I will first describe my improvements which are applicable to the receiving instruments.

My first improvement consists in automatically tapping or disturbing the powder in the sensitive tube, or in shaking the imperfect contact, so that immediately the electrical stimulus from the transmitter has ceased, the tube or imperfect contact regains its ordinary non-conductive state. This part of my invention is illustrated in figure 2, in which *j* represents the

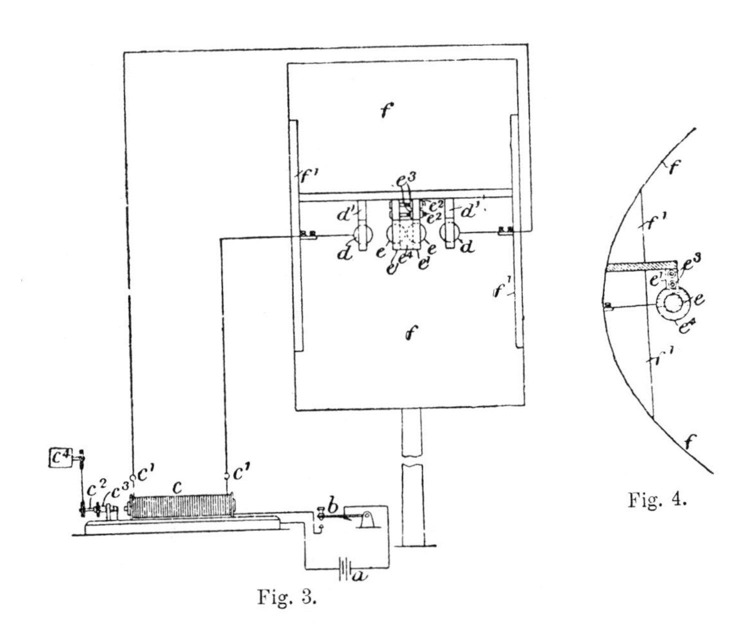

Fig. 3.

Fig. 4.

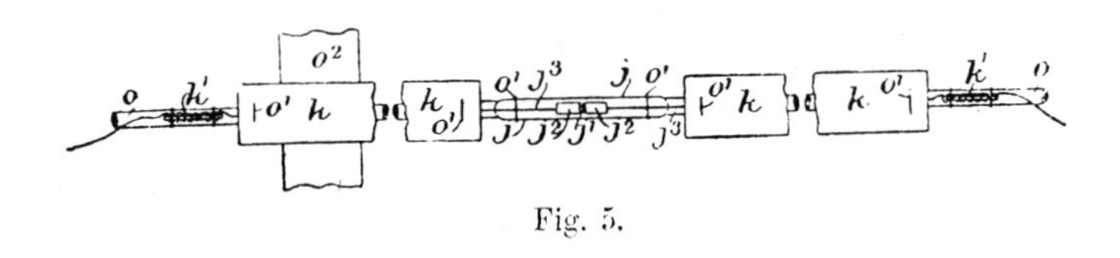

Fig. 5.

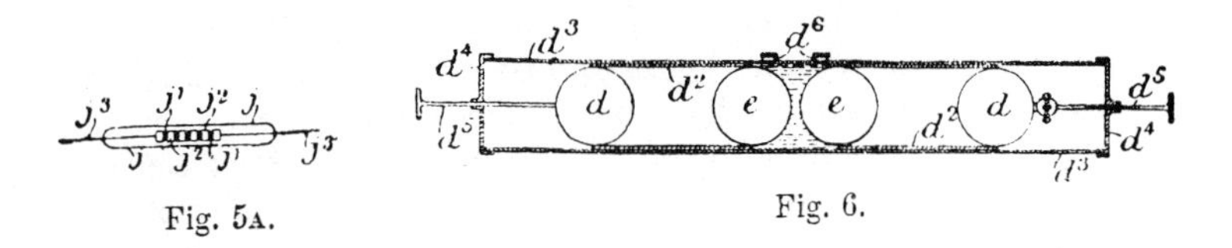

Fig. 5A.

Fig. 6.

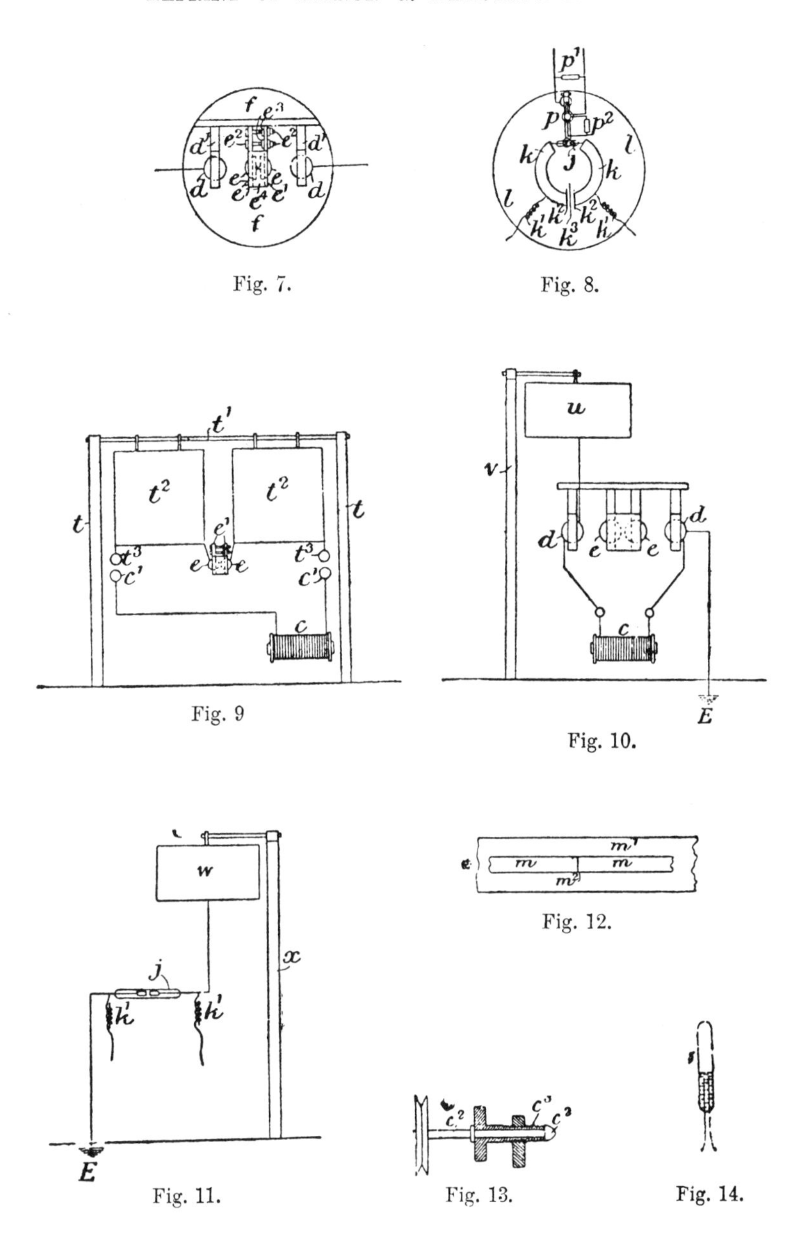

Fig. 7.

Fig. 8.

Fig. 9

Fig. 10.

Fig. 11.

Fig. 12.

Fig. 13.

Fig. 14.

sensitive tube and *p* the trembler or tapper. The current which flows through the sensitive tube or contact, and which is commenced under the influence of the electrical oscillations from the transmitting instrument, is allowed to actuate (directly, or indirectly by means of a relay) the trembler, which is similar to an electric bell. This trembler must be so arranged, as hereinafter explained, that the effect of the sparking at its vibrating contacts, and the jerky currents caused by self-induction, &c., are neutralised or removed.

The small hammer of the trembler hits the tube or imperfect contact and so stops the current, and consequently its own movement, which had been generated by the said current; and by this means the current automatically and almost instantaneously interrupts itself until another oscillation from the transmitting instrument again makes the sensitive tube or imperfect contact a conductor.

I find, however, that the current which can be started by the sensitive tube or contact is not sufficiently strong to work an ordinary trembler and receiving instrument.

To overcome this difficulty, instead of obliging the current of the circuit which contains the sensitive tube or contact to work the trembler and instrument, I use the said current for working a sensitive relay *n* (figure 2), which closes and opens the circuit of a stronger battery *r*, preferably of the Leclanché type. This current, which is much stronger than the current which runs through the sensitive tube or contact, works the trembler and other instruments. To prevent the sparks and jerks of current which would be caused by the self-induction of the relay from interfering with the action of the receiver, certain means must be taken similar to those referred to above in reference to the trembler or tapper, which will be explained hereafter. In the apparatus I have made I have found that the relay *n* should be one possessing small self-induction, and wound to a resistance of about 1000 ohms. It should preferably be able to work regularly with a current of a milliampère or less. The trembler or tapper *p* on the circuit of the relay *n* is similar in construction to that of a small electric bell, but having a shorter arm. I have used a trembler wound to 1000 ohms resistance, having a core of good soft iron hollow and split lengthways like most electro-magnets used in telegraph instruments.

The trembler must be carefully adjusted. Preferably the blows should be directed slightly upwards, so as to prevent the filings from getting caked. In place of tapping the tube the powder can be disturbed by slightly moving outwards and inwards one or both of the stops of the sensitive tube (see figure 5, j^1 j^2), the trembler *p* (figure 2) being replaced by a small electro-magnet or magnets or vibrator whose armature is connected to the stop.

I ordinarily work the receiving instrument *h*, which may be of any description, by a derivation as shown from the circuit, which works the trembler *p*. It can also, however, be worked in series with the trembler.

It is desirable that the receiving instrument, if on a derivation of the circuit which includes the trembler or tapper, should preferably have a resistance equal to the resistance of the trembler *p*.

A further improvement consists in the mode of construction of the sensitive tube.

I have noticed that a sensitive tube or imperfect contact, such as is shown in figure 1 T, is not perfectly reliable.

My tube as shown in figure 5 is, if carefully constructed, absolutely reliable, and by means of it the relay and trembler &c., can be worked with regularity like any other ordinary telegraphic instrument.

In figure 5, *j* is the sensitive tube containing two metallic plugs j^2 connected to the battery circuit, between which is placed powder of a conductive material j^1. The two plugs should preferably be made of silver, or may be two short pieces of thick silver wire of the same diameter as the internal diameter of the tube *j*, so as to fit tightly in it. The plugs j^2 j^2 are joined to two pieces of platinum wire j^3. The tube is closed and sealed on to the platinum wires j^3 at both ends. Many metals can be employed for producing the powder or filings j^1, but I prefer to use a mixture of two or more different metals. I find hard nickel to be the best metal, and I prefer to add to the nickel filings about four per cent of hard silver filings, which increase greatly the sensitiveness of the tube to electric oscillations. By increasing the proportion of silver powder or grains the sensitiveness of the tube also increases; but it is better for ordinary work not to use a tube of too great sensitiveness, as it might be influenced by atmospheric or other electricity.

The sensitiveness can also be increased by adding a very small amount of mercury to the filings and mixing up until the mercury is absorbed. The mercury must not be in such a quantity as to clot or cake the filings : an almost imperceptible globule is sufficient for a tube. Instead of mixing the mercury with the powder, one can obtain the same effects by slightly amalgamating the inner surfaces of the plugs which are to be in contact with the filings. Very little mercury must be used, just sufficient to brighten the surface of the metallic plugs without showing any free mercury or globules.

The size of the tube and the distance between the two metallic stops or plugs may vary under certain limits : the greater the space allowed for the powder, the larger or coarser ought to be the filings or grains.

I prefer to make my sensitive tubes of the following size—the tube j is $1\frac{1}{2}$ inch long and $\frac{1}{10}$ or $\frac{1}{12}$ of an inch internal diameter. The length of the stops j^2 is about $\frac{1}{5}$ of an inch, and the distance between the stops or plugs $j^2 j^2$ is about $\frac{1}{30}$ of an inch.

I find that the smaller or narrower the space is between the plugs in the tube, the more sensitive it proves ; but the space cannot under ordinary circumstances be excessively shortened without injuring the fidelity of the transmission.

Care must be taken that the plugs $j^2 j^2$ fit the tube exactly, as otherwise the filings might escape from the space between the stops, which would soon destroy the action of the sensitive tube.

The metallic powders ought not to be fine, but rather coarse, as can be produced by a large and rough file.

The powder should preferably be of uniform grain or thickness.

All the very fine powder or the excessively coarse powder ought to be removed from it by blowing or sifting.

It is also desirable that the powder or grains should be dry and free from grease or dirt, and the files used in producing the same ought to be frequently washed and dried, and used when warm.

The powder ought not to be compressed between the plugs,

but rather loose, and in such a condition that when the tube is tapped the powder may be seen to move freely.

The tube *j* may be sealed, but a vacuum inside it is not essential, except the slight vacuum which results from having heated it while sealing it. Care should also be taken not to heat the tube too much in the centre when sealing it, as it would oxidise the surfaces of the silver stops, and also the powder, which would diminish its sensitiveness. I have used, in sealing the tubes, a hydrogen and air flame.

A vacuum is, however, desirable, and I have used one of about $\frac{1}{1000}$ of an atmosphere obtained by a mercury pump.

In this case a small glass tube must be joined to a side of the tube *j* (figure 5), which is put in communication with the pump and afterwards sealed in the ordinary manner.

If the sensitive tube has been well made, it should be sensitive to the inductive effect of an ordinary electric bell when the same is working from one to two yards from the tube.

A sensitive tube well prepared should also instantly interrupt the current passing through it at the slightest tap or shake, provided it is inserted in a circuit in which there is little self-induction and small electro-motive force, such as a single cell.

In order to keep the sensitive tube *j* in good working order it is desirable, but not absolutely necessary, not to allow more than one milliampère to flow through it when active.

If a stronger current is necessary, several tubes may be put in parallel, provided they all get shaken by the tapper or trembler; but this arrangement is not always quite as satisfactory as the single tube.

It is preferable, when using sensitive tubes of the type I have described, not to insert in the circuit with it more than one cell of the Leclanché type, as a higher electro-motive force than 1·5 volts is apt to pass a current through the tube, even when no oscillations are transmitted.

I can, however, construct sensitive tubes capable of working with a higher electro-motive force.

Fig. 5A shows one of these tubes. In this tube, instead of one space or gap filled with filings, there are several spaces j^1 j^1, separated by plugs of tight-fitting silver wire. A tube thus constructed—observing also the rules of construction of my tubes in general—will work satisfactorily if the electro-motive force of the battery in circuit with the tube is equal to about 1·2 volts multiplied by the number of gaps.

With this tube also it is well not to allow a current of more than one milliampère to pass through it.

Figure 5 also shows the plates *k k*, which are joined to each end of the sensitive tube, and which correspond to the plates M N in figure 1.

The plates *k* (figure 5) are of copper or other metal, about half an inch or more broad, and may be about $\frac{1}{60}$ of an inch thick, and preferably of such a length as to be electrically tuned with the length of the wave of the electrical oscillations transmitted.

The means I adopt for fixing the proper length of the plates *k k* is as follows: I stick a rectangular strip of tinfoil (see figure 12) *m* about 20 inches long (the length depends on the supposed length of the wave that one is measuring), by means of a weak solution of gum, on to a glass plate m^1 (figure 12); then by means of a very sharp penknife or point and ruler I cut across the middle of the tinfoil, leaving a mark of division m^2. If this glass plate is held a few feet away from the origin of the electrical disturbances, and in such a position that the strips of tinfoil are about parallel to the line joining the centres of the two spheres in the transmitting apparatus, sparks will jump from one strip to the other at m^2. When the length of the strips of tinfoil *m* has been so adjusted as to approximate to the length of wave emitted from the oscillator, the sparking will occur at a greater distance from the oscillation producer when the strips are of suitable length. By shortening or lengthening the strips, therefore, it is easy to find the length most appropriate to the length of wave emitted by the oscillation producer. The length so found is the proper length for the plates *k*, or rather these should be about half an inch shorter on account of the length of the sensitive tube *j* (figure 5) connected between them.

The plates *k*, tube *j*, &c., are fastened to a thin glass tube *o*, preferably not longer than 12 inches, firmly fixed at one end to a firm piece of wood o^2, or the sensitive tube *j* may be fixed firmly at both ends—*i.e.*, preferably grasped near the ends of the tube containing the powder, and not at the ends of the tube *o o*, which serves as support.

By means of a tube with multiple gaps, as shown in figure 5A, it is also possible to work the trembler and also the signalling or other apparatus direct on the circuit which contains the sensitive tube, but I prefer when possible to work with the single-gap tube and the relay as shown. With a sensitive and specially constructed trembler it is also possible to work the trembler with the single-gap tube in series with it without a relay.

In order to increase the distance at which the receiver can be actuated by the radiation from the transmitter, I place the receiver (*i.e.*, the sensitive tube and plates) in the focal line of a cylindrical parabolic reflector *l* (figure 2), preferably of copper, and directed towards the transmitting station.

In determining the proper length of the plates of the receiver by means of the detector shown in figure 12, it is desirable to try the detector in the focus or focal line of the reflector, because the length of the strips or plates which gives the best result in a reflector differs slightly from the length which gives the best results without reflectors.

The reflector *l* (figure 2) should be preferably in length and opening not less than double the length of wave emitted from the transmitting instrument.

It is slightly advantageous for the focal distance of the reflector to be equal to one-fourth or three-fourths of the wavelength of the oscillation transmitted.

The plates *k* (figure 2) may be replaced by tubes or other forms of conductors.

A further improvement has for its object to prevent the electrical disturbances which are set up by the trembler and other apparatus in proximity or in circuit with the tube from themselves restoring the conductivity of the sensitive tube immediately after the trembler has destroyed it, as has been described.

This I effect by introducing into the circuits at the places marked p^1, p^2, q, h^1, in figure 2 high resistances having as little self-induction as possible. The action of the high resistances is that, while preventing any appreciable quantity of the current from passing through them when the apparatus is working, they nevertheless afford an easy path for the currents of high tension which would be formed at the moment when the circuit is broken, and thus prevent sparking at contacts or sudden jerks of currents, which would restore or maintain the conductivity of the sensitive tube.

These coils may conveniently be made by winding the wire (preferably of platinoid) on the bight, as it is sometimes termed, or double wound, to prevent them producing self-induction.

In figure 2, p^2 is one of these resistance coils which is inserted in a circuit connecting the vibrating contacts of the trembler p. I have used in the apparatus a coil which had a resistance about four times the resistance of the trembler p.

p^1 represents a similar resistance (also of about four times the resistance of the trembler) inserted in parallel across the terminals of the trembler.

A similar resistance q, figure 2, is placed in parallel on the terminals of the relay n (*i.e.*, the terminals which are connected to the circuit containing the sensitive tube).

The coil q should preferably have a resistance of about three or four times the resistance of the relay.

A similar resistance h^1 of about four times the resistance of the instrument is inserted in parallel across the terminals of the instrument.

In parallel across the terminals of the relay (*i.e.*, corresponding to the circuit worked by the relay) it is well to have a liquid resistance s constituted of a series of tubes, one of which is shown full size in figure 14 partially filled with water acidulated with sulphuric acid. The number of these tubes in series across the said terminals ought to be about ten for a circuit of 15 volts, so as to prevent, in consequence of their counter electro-motive force, the current of the local battery from passing through them, but allowing the high tension jerk of current generated at the opening of the circuit in the relay to pass smoothly across them without producing perturbating sparks at the movable contact of the relay.

A double-wound platinoid resistance may be used instead of the water resistance, provided its resistance be about 20,000 ohms.

A resistance similar to h should be inserted in parallel on the terminals of any apparatus or resistance which may be apt to give self-induction and which is near or connected to the receiver.

Condensers of suitable capacity may be substituted for the above-mentioned coils, but I prefer using coils or water resistances.

Another improvement has for its object to prevent the high frequency oscillations set up across the plates of the receiver by the transmitting instrument, which should pass through the sensitive tube, from running round the local battery wires and thereby weakening their effect on the sensitive tube or contact.

This I effect by connecting the battery wires to the sensitive tube or contact, or to the plates attached to the tube through small coils (see k^1 in the figures) possessing self-induction, which may be called choking coils, formed by winding in the ordinary manner a short length (about a yard) of thin and well-insulated wire round a core (preferably containing iron) two or three inches long.

Another improvement consists in a modified form of the plates connected to the sensitive tube, in order to make it possible to mount the receiver in an ordinary circular parabolic reflector. This part of my invention is illustrated in figure 8, in which *l* is an ordinary concave reflector. In this case the plates *k k* are curved and connected at one end to the sensitive tube *j*, and at the other to a small condenser formed by two metallic plates k^2 of about one inch square or more, facing each other with a very thin piece of insulating material k^3 between them. *p* is the trembler. The condenser may be omitted without much altering the effects obtainable.

The connection to the local circuit is made through two small choking coils k^1 k^1 as already described.

The adjustment of the whole is similar to that already described for the other receivers.

The receiver should be put in such a position as to intercept the reflected ring of radiations which exists behind or before the focus of the reflector, and ought to be preferably tuned with the length of wave of the oscillation transmitted, in similar manner to that before described, except that a ring of tinfoil with a single cut through it is employed.

I will now describe my improvements which are applicable to the transmitting instruments.

My first improvement consists in employing four spheres for producing the electrical oscillations.

This part of my invention is illustrated in figure 3, *d d*, *e e*, and in figure 6, *d d*, *e e*. The spheres *d d*, figure 3, are connected to the terminals c^1 of the secondary circuit of the induction coil *c*. The spheres *d d* are carried by insulating supports d^1 d^1.

Preferably the supports d^1 consist of plates of ebonite having holes to receive the balls, which are fixed by heating them sufficiently to fuse the ebonite and then holding them in place until they cool. *e e* are two similar balls on supports e^1 e^1, whose distance apart can be adjusted by ebonite bolts and nuts e^2 e^2 acting against other nuts e^3. e^4 is a flexible membrane, preferably of parchment paper, glued to the supports e^1 and forming a vessel which is filled with dielectric liquid, preferably vaseline-oil slightly thickened with vaseline.

The oil or insulating liquid between the spheres *e e* increases the power of the radiation, and also enables one to obtain constant effects, which are not easily obtained if the oil is omitted.

The balls *d* and *e* are preferably of solid brass or copper, and the distance they should be apart depends on the quantity and electro-motive force of the electricity employed, the effect increasing with the distance (especially by increasing the distance between the spheres *d* and the spheres *e*) so long as the discharge passes freely. With an induction coil giving an ordinary 8-inch spark the distance between *e* and *e* should be from $\frac{1}{25}$ to $\frac{1}{30}$ of an inch, and the distance between *d* and *e* about one inch.

When it is desired that the signal should only be sent in one direction, I place the oscillation producer in the focus or focal line of a reflector directed to the receiving station.

f (figure 3) and *f* (figure 4) show the cylindrical parabolic reflector made by bending a metallic sheet, preferably of brass or copper to form, and fixing it to metallic or wooden ribs f^1 (figure 3).

Other conditions being equal, the larger the balls the greater is the distance at which it is possible to communicate. I have generally used balls of solid brass of 4 inches diameter, giving oscillations of 10 inches length of wave.

Instead of spheres, cylinders or ellipsoids, &c., may be employed.

Preferably the reflector applied to the transmitter ought to be in length and opening the double at least of the length of wave emitted from the oscillator.

If these conditions are satisfied, and with a suitable receiver, a transmitter furnished with spheres of 4 inches diameter connected to an induction coil giving a 10-inch spark will transmit signals to two miles or more.

If a very powerful source of electricity giving a very long spark be employed, it is preferable to divide the spark-gap between the central balls of the oscillator into several smaller gaps in series. This may be done by introducing between the big balls smaller ones (of about half an inch diameter) held in position by ebonite frames.

Figure 6 shows a more compact form of oscillation producer. In this each pair of balls *d* and *e* is fixed by heat or otherwise in the end of tubes d^2 of insulating material, such as ebonite or vulcanite. The tubes d^2 fit tightly in another similar tube d^3 having covers d^4, through which pass the rods d^5 connecting the balls *d* to the conductors. One (or both) of the rods d^5 is connected to the ball *d* by a ball-and-socket joint, and has a screw thread upon it working in a nut in the cover d^4. By turning the rod, therefore, the distance of the balls *e* apart can be adjusted. d^6 are holes in the tube d^3, through which the vaseline-oil can be introduced into the space between the balls *e*.

A further improvement consists in causing one of the contacts of the vibrating brake applied to the induction coil to revolve rapidly.

This improvement has for its object to maintain the platinum contacts of the interrupter in good working order, and to prevent them sticking, &c.

This part of my invention is illustrated in figure 3 (c^2, c^3, c^4).

I obtain this result by having a revolvable central core c^2 (figure 3 and figure 13) in the ordinary screw c^3, which is in communication with the platinum contacts. I cause the said central core with one of the platinum contacts attached to it to revolve by coupling it to a small electric motor c^4.

This motor can be worked by the same circuit that works the coil, or if necessary by a separate circuit—the connections are not shown in the drawing.

By this means the regularity and power of the discharge of an ordinary induction coil with a trembler brake are greatly improved.

The induction coil *c* (figure 3) may, however, be replaced by any other source of high-tension electricity.

When working with large amounts of energy it is, however, better to keep the coil of the transformer constantly working for the time during which one is transmitting, and, instead of interrupting the current of the primary, interrupting the discharge of the secondary.

In this case the contacts of the key should be immersed in oil, as otherwise, owing to the length of the spark, the current will continue to pass after the contacts have been separated.

A further improvement has for its object to facilitate the focussing of the electric rays.

This part of my invention is illustrated in figure 7, in which a view is given of the modified oscillation producer mounted in the focus of an ordinary parabolic reflector *f*.

The oscillator in this case is different from the one I have previously described, because instead of being constituted of two spheres it is made of two hemispheres *e e* separated by a small space filled with oil or other dielectric. The spark between the hemispheres takes place in the dielectric from small projections at the centres of the hemispheres. The working and adjusting of this oscillator are similar to that of the one previously described.

This arrangement may be also solidly mounted in an ebonite tube, as shown in figure 6.

A receiver which may be used with this transmitter is shown in figure 8, and has already been described.

It is not essential to have a reflector at the transmitters and receivers, but in their absence the distance at which one can communicate is much smaller.

Figure 9 shows another modified form of transmitter with

which one can transmit signals to considerable distances without using reflectors.

In figure 9, *t t* are two poles connected by a rope t^1, to which are suspended by means of insulating suspenders two metallic plates t^2 t^2 connected to the spheres *e* (in oil, or other dielectric, as before) and to the other balls t^3 in proximity to the spheres c^1, which are in communication with the coil or transformer *c*. The balls t^3 are not absolutely necessary, as the plates t^2 may be made to communicate with the coil or transformer by means of thin insulated wires. The receiver I adopt with this transmitter is similar to it, except that the spheres *e* are replaced by the sensitive tube or imperfect contact *j* (figure 5), whilst the spheres t^3 may be replaced by the choking coils k^1 in communication with the local circuit. If a circular-tuned receiver of large size be employed, the plates t^2 may be omitted from the receiver. I have observed that, other conditions being equal, the larger the plates at the transmitter and receiver, and the higher they are from earth, and to a certain extent the farther apart they are, the greater is the distance at which correspondence is possible.

For permanent installations it is convenient to replace the plates by metallic cylinders closed at one end, and put over the pole like a hat, and resting on insulators. By this arrangement no wet can come to the insulators, and the effects obtainable are better in wet weather.

A cone or hemisphere may be used in place of a cylinder. The pole employed ought preferably to be dry and tarred.

Where obstacles, such as many houses or a hill or mountains, intervene between the transmitter and the receiver, I have devised and adopt the arrangement shown in figures 10 and 11.

In the transmitting instrument, figure 10, I connect one of the spheres *d* to earth E preferably by a thick wire, and the other to a plate or conductor *u*, which may be suspended on a pole *v* and insulated from earth. Or the spheres *d* may be omitted and one of the spheres *e* be connected to earth and the other to the plate or conductor *u*.

At the receiving station, figure 11, I connect one terminal of the sensitive tube or imperfect contact *j* to earth E, preferably also by a thick wire, and the other to a plate or conductor *w*, preferably similar to *u*. The plate *w* may be suspended on a pole *x*, and should be insulated from earth. The larger the plates of the receiver and transmitter, and the higher from the earth the plates are suspended, the greater is the distance at which it is possible to communicate at parity of other conditions.

The figure does not show the trembler or tapping arrangement. k^1 k^1 are the choking coils, which are connected to the battery circuit, as has been explained with reference to the previous figures.

At the receiver it is possible to pick up the oscillations from the earth or water without having the plate *w*. This may be done by connecting the terminals of the sensitive tube *j* to two earths, preferably at a certain distance from each other and in a line with the direction from which the oscillations are coming. These connections must not be entirely conductive, but must contain a condenser of suitable capacity, say of one square yard surface (parafined paper as dielectric).

Balloons can also be used instead of plates on poles, provided they carry up a plate or are themselves made conductive by being covered with tinfoil. As the height to which they may be sent is great, the distance at which communication is possible becomes greatly multiplied. Kites may also be successfully employed if made conductive by means of tinfoil.

When working the described apparatus, it is necessary either that the local transmitter and receiver at each station should be at a considerable distance from each other, or that they should be screened from each other by metal plates. It is sufficient to have all the telegraphic apparatus in a metal box (except the reading instrument), and any exposed part of the circuit of the receiver enclosed in metallic tubes which are in electrical communication with the box (of course the part of the apparatus which has to receive the radiation from the distant station must not be enclosed, but possibly screened from the local transmitting instrument by means of metallic sheets).

When the apparatus is connected to the earth or water the receiver must be switched out of circuit when the local transmitter is at work, and this may also be done when the apparatus is not earthed.

Having now particularly described and ascertained the nature of my said invention, and in what manner the same is to be performed, I declare that what I claim is—

1. The method of transmitting signals by means of electrical impulses to a receiver having a sensitive tube or other sensitive form of imperfect contact capable of being restored with certainty and regularity to its normal condition substantially as described.

2. A receiving instrument consisting of a sensitive imperfect contact or contacts, a circuit through the contact or contacts, and means for restoring the contact or contacts, with certainty and regularity, to its or their normal condition after the receipt of an impulse substantially as described.

3. A receiving instrument consisting of a sensitive imperfect contact or contacts, a circuit through the contact or contacts, and means actuated by the circuit for restoring with certainty and regularity the contact or contacts to its or their normal condition after the receipt of an impulse.

4. In a receiving instrument such as is mentioned in claims 2 and 3, the use of resistances possessing low self-induction, or other appliances for preventing the formation of sparks at contacts or other perturbating effects.

5. The combination with the receivers such as are mentioned in claims 2 and 3 of resistances or other appliances for preventing the self-induction of the receiver from affecting the sensitive contact or contacts substantially as described.

6. The combination with receivers such as herein above referred to of choking coils substantially as described.

7. In receiving instruments consisting of an imperfect contact or contacts sensitive to electrical impulses, the use of automatically working devices for the purpose of restoring the contact or contacts with certainty and regularity to their normal condition after the receipt of an impulse substantially as herein described.

8. Constructing a sensitive non-conductor capable of being made a conductor by electrical impulses of two metal plugs or their equivalents, and confining between them some substance such as described.

9. A sensitive tube containing a mixture of two or more powders, grains, or filings, substantially as described.

10. The use of mercury in sensitive imperfect electrical contacts substantially as described.

11. A receiving instrument having a local circuit, including a sensitive imperfect electrical contact or contacts, and a relay operating an instrument for producing signals, actions, or manifestations substantially as described.

12. Sensitive contacts in which a column of powder or filings (or their equivalent) is divided into sections by means of metallic stops or plugs substantially as described.

13. Receivers substantially as described and shown in figures 5 and 8.

14. Transmitters substantially as described and shown at figures 6 and 7.

15. A receiver consisting of a sensitive tube or other imperfect contact inserted in a circuit, one end of the sensitive tube or other imperfect contact being put to earth whilst the other end is connected to an insulated conductor.

16. The combination of a transmitter having one end of its sparking appliance or poles connected to earth, and the other to an insulated conductor, with a receiver as is mentioned in claim 15.

17. A receiver consisting of a sensitive tube or other imperfect contact inserted in a circuit, and earth connections to each end of the sensitive contact or tube through condensers or their equivalent.

18. The modifications in the transmitters and receivers, in which the suspended plates are replaced by cylinders or the like placed hat-wise on poles, or by balloons or kites substantially as described.

19. An induction coil having a revolving make and break substantially as and for the purposes described.

Dated this 2nd day of March 1897.

GUGLIELMO MARCONI.

INDEX.

THE END.

PRINTED BY WILLIAM BLACKWOOD AND SONS.

HISTORY OF BROADCASTING:
Radio To Television
An Arno Press/New York Times Collection

Archer, Gleason L.
Big Business and Radio. 1939.

Archer, Gleason L.
History of Radio to 1926. 1938.

Arnheim, Rudolf.
Radio. 1936.

Blacklisting: Two Key Documents. 1952–1956.

Cantril, Hadley and Gordon W. Allport.
The Psychology of Radio. 1935.

Codel, Martin, editor.
Radio and Its Future. 1930.

Cooper, Isabella M.
Bibliography on Educational Broadcasting. 1942.

Dinsdale, Alfred.
First Principles of Television. 1932.

Dunlap, Orrin E., Jr.
Marconi: The Man and His Wireless. 1938.

Dunlap, Orrin E., Jr.
The Outlook for Television. 1932.

Fahie, J. J.
A History of Wireless Telegraphy. 1901.

Federal Communications Commission.
Annual Reports of the Federal Communications Commission. 1934/1935–1955.

Federal Radio Commission.
Annual Reports of the Federal Radio Commission. 1927–1933.

Frost, S. E., Jr.
Education's Own Stations. 1937.

Grandin, Thomas.
The Political Use of the Radio. 1939.

Harlow, Alvin.
Old Wires and New Waves. 1936.

Hettinger, Herman S.
A Decade of Radio Advertising. 1933.

Huth, Arno.
Radio Today: The Present State of Broadcasting. 1942.

Jome, Hiram L.
Economics of the Radio Industry. 1925.

Lazarsfeld, Paul F.
Radio and the Printed Page. 1940.

Lumley, Frederick H.
Measurement in Radio. 1934.

Maclaurin, W. Rupert.
Invention and Innovation in the Radio Industry. 1949.

Radio: Selected A.A.P.S.S. Surveys. 1929–1941.

Rose, Cornelia B., Jr.
National Policy for Radio Broadcasting. 1940.

Rothafel, Samuel L. and Raymond Francis Yates.
Broadcasting: Its New Day. 1925.

Schubert, Paul.
The Electric Word: The Rise of Radio. 1928.

Studies in the Control of Radio: Nos. 1–6. 1940–1948.

Summers, Harrison B., editor.
Radio Censorship. 1939.

Summers, Harrison B., editor.
A Thirty-Year History of Programs Carried on National Radio Networks in the United States, 1926–1956. 1958.

Waldrop, Frank C. and Joseph Borkin.
Television: A Struggle for Power. 1938.

White, Llewellyn.
The American Radio. 1947.

World Broadcast Advertising: Four Reports. 1930–1932.